Lecture Notes in Computer Science   9832

Commenced Publication in 1973
Founding and Former Series Editors:
Gerhard Goos, Juris Hartmanis, and Jan van Leeuwen

M. Elena Renda · Miroslav Bursa
Andreas Holzinger · Sami Khuri (Eds.)

Information Technology in Bio- and Medical Informatics

7th International Conference, ITBAM 2016
Porto, Portugal, September 5–8, 2016
Proceedings

Springer

Editors
M. Elena Renda
Institute of Informatics and Telematics
Pisa
Italy

Miroslav Bursa
Czech Technical University in Prague
Prague
Czech Republic

Andreas Holzinger
Medical University Graz
Graz
Austria

Sami Khuri
San José State University
San Jose, CA
USA

ISSN 0302-9743   ISSN 1611-3349   (electronic)
Lecture Notes in Computer Science
ISBN 978-3-319-43948-8   ISBN 978-3-319-43949-5   (eBook)
DOI 10.1007/978-3-319-43949-5

Library of Congress Control Number: 2016946948

LNCS Sublibrary: SL3 – Information Systems and Applications, incl. Internet/Web, and HCI

Printed on acid-free paper

This Springer imprint is published by Springer Nature
The registered company is Springer International Publishing AG Switzerland

Preface

Biomedical engineering and medical informatics represent challenging and rapidly growing areas. Applications of information technology in these areas are of paramount importance. Building on the success of ITBAM 2010, ITBAM 2011, ITBAM 2012, ITBAM 2013, ITBAM 2014, and ITBAM 2015, the aim of the seventh ITBAM conference was to continue bringing together scientists, researchers, and practitioners from different disciplines, namely, from mathematics, computer science, bioinformatics, biomedical engineering, medicine, biology, and different fields of life sciences, to present and discuss their research results in bioinformatics and medical informatics. We hope that ITBAM will serve as a platform for fruitful discussions between all attendees, where participants can exchange their recent results, identify future directions and challenges, initiate possible collaborative research, and develop common languages for solving problems in the realm of biomedical engineering, bioinformatics, and medical informatics. The importance of computer-aided diagnosis and therapy continues to draw attention worldwide and has laid the foundations for modern medicine with excellent potential for promising applications in a variety of fields, such as telemedicine, Web-based healthcare, analysis of genetic information, and personalized medicine.

Following a thorough peer-review process, we selected nine long papers for oral presentation and 11 short papers for poster session for the seventh annual ITBAM conference (from a total of 26 contributions). The organizing committee would like to thank the reviewers for their excellent job. The articles can be found in the proceedings and are divided to the following sections: *Biomedical Data Analysis and Warehousing, Information Technologies in Brain Sciences,* and *Social Networks and Process Analysis in Biomedicine.* The papers show how broad the spectrum of topics in applications of information technology to biomedical engineering and medical informatics is.

The editors would like to thank all the participants for their high-quality contributions and Springer for publishing the proceedings of this conference. Once again, our special thanks go to Gabriela Wagner for her hard work on various aspects of this event.

June 2016

M. Elena Renda
Miroslav Bursa
Andreas Holzinger
Sami Khuri

Organization

General Chair

Christian Böhm — University of Munich, Germany

Program Committee Co-chairs

Miroslav Bursa	Czech Technical University, Czech Republic
Andreas Holzinger	Medical University Graz, Austria
Sami Khuri	San José State University, USA
M. Elena Renda	IIT - CNR, Pisa, Italy (Honorary Chair)

Program Committee

Tatsuya Akutsu	Kyoto University, Japan
Andreas Albrecht	Queen's University Belfast, Ireland
Peter Baumann	Jacobs University Bremen, Germany
Miroslav Bursa	Czech Technical University, Czech Republic
Christian Böhm	University of Munich, Germany
Rita Casadio	University of Bologna, Italy
Sònia Casillas	Universitat Autònoma de Barcelona, Spain
Kun-Mao Chao	National Taiwan University, Taiwan
Vaclav Chudacek	Czech Technical University, Czech Republic
Hans-Dieter Ehrich	Technical University of Braunschweig, Germany
Christoph M. Friedrich	University of Applied Sciences Dortmund, Germany
Jan Havlik	Czech Technical University, Czech Republic
Volker Heun	Ludwig-Maximilians-Universität München, Germany
Andreas Holzinger	Medical University Graz, Austria
Larisa Ismailova	NRNU MEPhI, Moscow, Russia
Alastair Kerr	University of Edinburgh, UK
Sami Khuri	San Jose State University, USA
Jakub Kuzilek	Czech Technical University, Czech Republic
Lenka Lhotska	Czech Technical University, Czech Republic
Roger Marshall	Plymouth State University, USA
Elio Masciari	ICAR-CNR, Università della Calabria, Italy
Nadia Pisanti	University of Pisa, Italy
Cinzia Pizzi	Università degli Studi di Padova, Italy
Clara Pizzuti	ICAR-CNR, Italy
Maria Elena Renda	CNR-IIT, Italy
Stefano Rovetta	University of Genova, Italy
Roberto Santana	University of the Basque Country (UPV/EHU), Spain

Huseyin Seker	De Montfort University, UK
Jiri Spilka	Czech Technical University, Czech Republic
Kathleen Steinhofel	King's College London, UK
Songmao Zhang	Chinese Academy of Sciences, China
Qiang Zhu	The University of Michigan, USA

Contents

Poster Session

Biomedical Data Analysis and Warehousing

What Do the Data Say in 10 Years of Pneumonia Victims?
A Geo-Spatial Data Analytics Perspective

Maribel Yasmina Santos[1]([⊠]), António Carvalheira Santos[2],
and Artur Teles de Araújo[2]

[1] ALGORITMI Research Centre, University of Minho, Guimarães, Portugal
maribel@dsi.uminho.pt
[2] Portuguese Lung Foundation, Lisboa, Portugal
antonio.carvalheira@gmail.com,
artur@telesdearaujo.com

Abstract. The need to integrate, store, process and analyse data is continuously growing as information technologies facilitate the collection of vast amounts of data. These data can be in different repositories, have different data formats and present data quality issues, requiring the adoption of appropriate strategies for data cleaning, integration and storage. After that, suitable data analytics and visualization mechanisms can be used for the analysis of the available data and for the identification of relevant knowledge that support the decision-making process. This paper presents a data analytics perspective over 10 years of pneumonia incidence in Portugal, pointing the evolution and characterization of the mortal victims of this disease. The available data about the individuals was complemented with statistical data of the country, in order to characterize the overall incidence of this disease, following a spatial analysis and visualization perspective that is supported by several analytical dashboards.

Keywords: Business intelligence · (Spatial) data warehouse · Data analytics · Pneumonia

1 Introduction

Business intelligence and analytics have become increasingly relevant over the past two decades, reflecting the magnitude and impact of data-related problems [1]. This is a field of knowledge that has been using data warehouses as data repositories, providing an integrated and homogeneous set of data used in analytical contexts to support the decision making process [2]. A data warehouse can then be analysed using different supporting technologies as on-line analytical processing [3] or data mining algorithms [4], among others.

When data includes spatial attributes, like locations, the data model of a data warehouse can include spatial dimensions or attributes, allowing the analysis of the available data under this spatial perspective. Data warehouses with spatial characteristics have also become a topic of growing interest in recent years [5], being their logical design based on the multidimensional model, providing support for the definition of spatial data

© Springer International Publishing Switzerland 2016
M.E. Renda et al. (Eds.): ITBAM 2016, LNCS 9832, pp. 3–21, 2016.
DOI: 10.1007/978-3-319-43949-5_1

dimensions and/or spatial measures. Dimensions represent the analysis axes, while measures are the variables being analysed against the different dimensions. The implementation of spatial On-Line Analytical Processing (OLAP) tools can be achieved through solutions that are OLAP dominant, Geographical Information Systems (GIS) dominant, or both in a mixed solution [6]. Those tools are powerful decision-making instruments as they allow users to explore and analyse data in user-friendly applications and to formulate *ah-doc* queries on these data.

This paper presents a data analytics perspective using the data available in a data warehouse, with spatial characteristics, integrating data related with the incidence of pneumonia in Portugal, from 2002 to 2011, integrating 369 160 records. Besides these data, with the characterization of the affected individuals and other related pathologies, it was possible to integrate statistical data collected in the last census exercise undertaken in Portugal in 2011 [7].

The work here presented shows how several dashboards with spatial data, implemented over the mentioned data warehouse, were used in a data-driven analytical approach for an interactive analysis of the data, highlighting valuable information to characterize the incidence of a disease that, for respiratory infections, is the leading cause of death and hospital admissions in Portugal [8], following a global trend, as stated by the World Health Organization, mentioning that the lower respiratory infections are among the 10 leading causes of death at a Mundial level [9].

This paper is organized as follows. Section 2 presents related work. Section 3 summarizes the adopted methodology. Section 4 describes the data available for analysis. Section 5 summarizes some of the main findings in understanding pneumonia fatalities. Section 6 concludes with some remarks about the described work and guidelines for future work.

2   Related Work

Several works in the literature show the analysis of data about respiratory diseases, and some of them about pneumonia, following data analysis strategies that try to point out tendencies, patterns or models that can be useful in the decision-making process. Some of these works use statistical approaches, or techniques usually used in business intelligence contexts like OLAP or data mining. Although with relevant contributions to the community, none of these works was able to integrate such vast volume of data, providing a comprehensive knowledge about the incidence of this disease and, more important, its fatalities. This is of upmost importance for decision-makers in the definition of adequate actions to fight this disease.

The work of [10] presents a descriptive analysis of data retrieved from the medical reports at the Tawau General Hospital in Malaysia, where patients filled a special form that required information such as the patient age, area of origin, parent's smoking background, parent's medical background (if known), patient medical background (if known), among other relevant information. The performed analyses identified the profile of the patients who were admitted to this hospital. The authors report that there are several factors that may have caused the pneumonia, such as family background, or genetic and environmental factors, alerting the government authorities and doctors for the need of taking appropriate actions. In total, data from 102 patients were used in this study. As main results, the authors point that 86.27 % of the patients are from rural areas, underlining poor hygiene as an important factor in the origin of pneumonia in Malaysia.

With a higher number of studied individuals, the work of [11] reported that pneumonia is a disease most often fatal, which can be acquired by patients during their stay in intensive care units. Data from patients admitted to the intensive care unit at the Friedrich Schiller University Jena were collected and stored in a real-time database, totalizing 11 726 cases in two years. Based on these, the authors developed an early warning system for the onset of pneumonia that combines Alternating Decision Trees for supervised learning and Sequential Pattern Mining. The implemented detection system estimates a prognosis of pneumonia every 12 h for each patient. In case of a positive prognosis, an alert is generated. In this case, data mining algorithms, one of the data analysis techniques used by business intelligence systems, showed to be useful in the analysis of the collected data.

In [12], the authors show a study that allowed the development and validation of an ALI (Acute Lung Injury) prediction score in a population-based sample of patients at risk. For the prediction score, the authors used a logistic regression analysis. Patients at risk of acquiring an acute respiratory distress syndrome, the most severe form of ALI, were first identified in an electronic alert system that uses a Microsoft SQL-based database and a data mart for storing data about patients in an intensive care unit. A total of 876 records were analyzed, divided in 409 patients for the retrospective derivation cohort and 467 for the validation cohort.

More recently, [13] proposed the use of Disjunctive Normal Forms for predicting hospital and 90-day mortality from instance-based patient data, comprising demographic, genetic, and physiologic information in a cohort of patients admitted with severe acquired pneumonia. The authors developed two algorithms for learning Disjunctive Normal Forms, which make available a set of rules that map data to the outcome of interest. The authors show that Disjunctive Normal Forms achieve higher prediction performance quality when compared to a set of state-of-the-art machine learning models. Regarding data, patients with community-acquired pneumonia, a common cause of sepsis, were recruited as part of a study conducted in the United States (Western Pennsylvania, Connecticut, Michigan, and Tennessee) between November 2001–November 2003. Eligible subjects had 18 or more years old and had a clinical and radiologic diagnosis of pneumonia. Among the 2 320 patients enrolled, the authors restricted their analysis to 1 815 individuals admitted to the hospital.

3 Methodological Approach

The analysis of vast amounts of data with the aim of identifying useful patterns or insights can be achieved following an exploratory data analysis approach, which aims identifying relationships between different variables that seem interesting, checking if there is any evidence for or against a stating hypothesis [14]. In this process, it is very important looking for problems in the available data, as well as identifying complementary data that could add value to the data under analysis. In this sense, exploratory

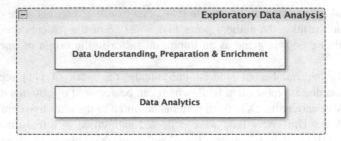

Fig. 1. Exploratory data analysis (different roles)

data analysis is useful in a preliminary analysis of the data, in order to understand, prepare and enrich it, and later, for the analysis itself in the data analytics approach, supporting the decision making process (Fig. 1).

Starting with the data understanding, preparation and enrichment, this allows the enhancement of a data set for data analysis purposes. In our previous work [7], it was possible to do an extensive analysis of the data, in order to get a deep knowledge about it, analyzing the available attributes, verifying all possible values, identifying data quality problems, enriching the data with external data sources, modeling the analytical repository for storing the data for analysis and, finally, implementing that repository. All these stages iteratively add value to the initial collected data, either cleaning the data (removing errors or problems) or completing it with additional sources (sometimes external to the organizations). For the concretization of such an analytical data repository, Fig. 2 summarizes the main followed steps, some of them possible through exploratory data analysis.

```
┌─ Data Understanding, Preparation & Enrichment ┐
│  ┌─────────────────────────────────────────┐  │
│  │         Data Understanding              │  │
│  └─────────────────────────────────────────┘  │
│  ┌─ Data Preparation ──────────────────────┐  │
│  │  ┌───────────────────────────────────┐  │  │
│  │  │         Data Cleaning             │  │  │
│  │  └───────────────────────────────────┘  │  │
│  │  ┌───────────────────────────────────┐  │  │
│  │  │        Data Derivation            │  │  │
│  │  └───────────────────────────────────┘  │  │
│  │  ┌───────────────────────────────────┐  │  │
│  │  │        Data Integration           │  │  │
│  │  └───────────────────────────────────┘  │  │
│  └─────────────────────────────────────────┘  │
│  ┌─────────────────────────────────────────┐  │
│  │         Data Enrichment                 │  │
│  └─────────────────────────────────────────┘  │
│  ┌─────────────────────────────────────────┐  │
│  │         Data Storage                    │  │
│  └─────────────────────────────────────────┘  │
└───────────────────────────────────────────────┘
```

Fig. 2. Steps in the data understanding, preparation and enrichment

After the understanding, preparation and cleaning of the data, exploratory data analysis can be used for data analytics, making use of tables or specific charts or graphs to obtain useful insights on data. In this task, the user/researcher must do critical evaluations of the findings, identifying interesting paths for analysis and, also, those that do not worth pursuing, as data are not providing useful or enough evidence of results [14]. The overall goal is to show the data, summarizing the relevant evidences and identifying interesting patterns.

For data analytics with exploratory data analysis, this work makes use of analytical graphics (in this case with a geo-spatial focus), trying to make informative and useful data graphics [15, 16]. For Tufte [15], excellent graphics exemplify the deep fundamental principles of analytical design in action, mentioning 6 fundamental principles of the analytical design: 1. Show comparisons, contrasts, differences; 2. Causality, mechanism, structure, explanation; 3. Multivariate analysis; 4. Integration of evidence; 5. Documentation; and, 6. Content counts most of all (Fig. 3).

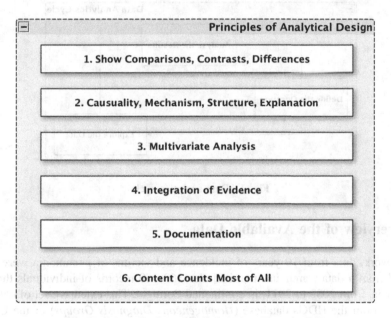

Fig. 3. Principles for analytical design (Source: [15])

Going through these principles, showing comparison is considered the basis of all scientific investigation, as showing evidence for a hypothesis is always relative to another competing hypothesis. Also, it is useful to show the causal framework when thinking about a question, meaning that data graphics could include information about possible causes, useful in suggesting hypotheses or refuting them. The most important is that this will raise new questions that can be followed up with new data analyses, which should be multivariate, as usually there are many attributes that can be measured or analyzed. Data graphics should attempt to show this information as much as possible, rather than reducing things down to one or two features. In those data graphics, numbers, words, images and diagrams can be included to tell a story, making use of many modes of data presentation and integrating as much evidence as possible. When describing and documenting the evidences, data graphics must be properly documented with labels, scales and sources, telling a completely story by itself, avoiding the need for extra texts or descriptions for interpreting a plot. For presenting the results, the content includes a good question, the approach for addressing it and the information that is necessary for answering that question [14].

All these principles of analytical design when included in data analytics through exploratory data analysis give support to the Data Analytics Cycle followed in this work, in which a question starts the cycle, being followed by data exploration. The analysis of results looks into the obtained findings in order to identify new questions or analytical paths for data analysis (Fig. 4).

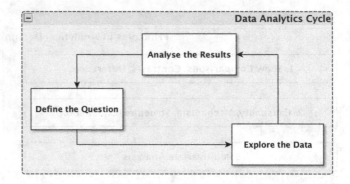

Fig. 4. Data analytics cycle

4 Overview of the Available Data

In this work, data from 10 years of incidence and victims of pneumonia were used, selected from a data warehouse that includes 369 169 records of individuals that had pneumonia, from 2002 to 2011, in continental Portugal. This extensive set of data was extracted from the HDGs database (*Homogeneous Diagnosis Groups*) of the Central Administration of Health Services - ACSS (*Administração Central dos Serviços de Saúde*). All the data, after an extensive work of extraction, transformation and loading, was stored in an analytical data repository now used for data analytics [7]. Besides the information of the individuals and their characteristics, this analytical repository also includes statistical data collected in the latest census exercise carried out in Portugal, in 2011 [17]. This will allow the verification of the most affected regions, regarding the number of mortal victims and the living population.

In our previous works [7, 18], the available data was analysed to characterize the disease and its evolution along the years. It was possible to verify that the consequences of the disease change depending on the age of the patients that are affected, on their

Table 1. Data attributes for analysis

Attribute	Description	Type	Values
Admission days	Total number of days in a healthcare facility	Integer	Min: 0, Max: 1032, Median: 8, Standard deviation: 11.7
Admission days class	Classes for the number of days in a healthcare facility	Categorical	[0–3], [4–6], [7–10], [11–29], [30+]
Age	Age of the patient	Integer	Min: 0, Max: 111, Median: 76, Standard deviation: 26.9
Age groups	Classes for the age of the patient	Categorical	[0–1], [2–5], [6–9], [10–13], [14–17], [18–34], [35–64], [65–79], [80+]
District	District of the patient	Categorical	18 Districts (Continental Portugal): Aveiro, Braga, Porto, Lisboa, Coimbra,…
Gender	Gender of the patient	Categorical	F (Female), M (Male)
Longitude	Longitude coordinate	Numeric	Min: −9.462, Max: −6.210
Latitude	Latitude coordinate	Numeric	Min: 37.000, Max: 42.140
Mortal victim	Flag that states if the patient was, or not, a mortal victim	Binary	0: Not a mortal victim 1: Mortal victim
Municipality	Municipality of the patient	Categorical	279 Municipalities of Continental Portugal
Number of residents	Number of residents in a given parish	Integer	Min: 31, Max: 66 250, Median: 820, Standard Deviation: 5 083
Parish	Parish of the patient	Categorical	3445 Parishes of Continental Portugal
Pneumonias counter	Event-tracking measure to summarize data	Integer	1
Readmissions number	Number of readmissions in a healthcare facility	Integer	Min: 0, Max: 13, Median: 0, Standard deviation: 0.63
Year	Year of the admission/visit to the healthcare facility	Integer	[2002–2011]

physical condition, as well as other pathologies that may affect the course of the disease. These studies have shown that the number of cases of pneumonia has increased 33.9 % in the decade under analysis and that the number of fatalities increased at a higher rate, reaching 65.3 % from 2002 to 2011 [7]. Moreover, it was possible to verify that a significant number of patients that died, as consequence of this disease, had a very short admission in the hospital, in terms of staying there for treatment. Regarding related pathologies, some patients with pneumonia also presented other diseases like the chronic pulmonary disease, the chronic cardiac disease, the chronic renal disease, the chronic pancreatic disease, the chronic hepatic disease, and the diabetes mellitus disease [18].

Having this preliminary knowledge about the incidence of the disease, this paper follows a data-driven analytics approach for a deepest analysis of a subset of the available data, trying to understand the course of the disease, in terms of fatalities, focusing in its geo-spatial incidence and in the identification of the more affected regions, considering several dimensions of analysis. With regard to location, it is important to mention that due to privacy concerns, the location where the patients' live/lived is associated with the centroids of the corresponding parishes and not to a specific street, for instance. To allow the proper visualization of the available information on a map, the centroids' coordinates were shacked in order to slightly distribute them in a map, around the corresponding parishes, showing the number of patients in each location. For the study presented in this paper, the relevant data attributes for analysis are summarized in Table 1, presenting the attribute name, description, type, and its possible values.

Before proceeding with the data analytics approach, let us briefly explore the available data in order to provide some background knowledge about the phenomena under analysis. Figure 5 shows two distribution graphs with the number of cases of pneumonia by year (Fig. 5(a)), and the number of cases by age (Fig. 5(b)). In the first case, it is possible to verify the increase that the disease has presented along these ten years. In the second, the incidence of cases increases substantially after the sixties, reaching the highest value in patients in the eighties. Also, as shown in the red area of Fig. 5(b), the number of mortal victims increases with age. Regarding the classes for the age, this is the first time that these specific ranges are used and the aim is to provide a deeper insight in these several groups.

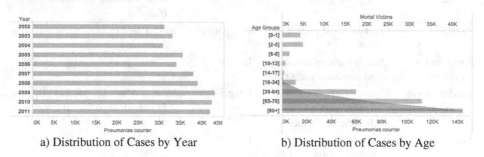

a) Distribution of Cases by Year   b) Distribution of Cases by Age

Fig. 5. Number of cases by year and age (Color figure online)

Patients with pneumonia can have shorter or longer stays in the healthcare facilities for treatment. In many cases, severe conditions require longer stays or, in some cases, very short stays are verified when the patients died because it was too late for treatment, for instance. As we can see in Fig. 6(a), very long stays, superior to 30 days, are mainly associated to individuals with more than forty years old, while shorter stays can be verified in all ages. This is better seen in the graph of Fig. 6(b), which depicts a smoothed colour density representation of a scatterplot, obtained through a kernel density estimate [19].

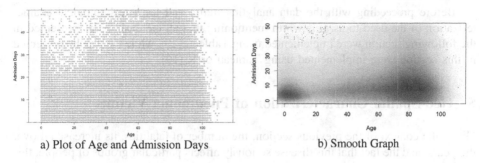

a) Plot of Age and Admission Days   b) Smooth Graph

Fig. 6. Analysis of ages and number of admission days

When we look into the relation between the age of the patients, the classes that were created for the number of days in the hospital, and if the patient is, or not, a mortal victim, the pattern previously mentioned emerges even stronger. For those that died as consequence of the disease, flag mortal victim equal to 1 in the right part of Fig. 7(a), the patients had an average age of approximately eighty years old, being this value very homogeneous for all the classes of admission days. In the case of patients that were not mortal victims, flag mortal victim equal to 0 in the left part of Fig. 7(a), stays in the healthcare facilities tend to be longer as age increases. The information obtained from Fig. 7(b) is very relevant as shows that, for a significant number of mortal victims, shorter stays in the hospital were verified, meaning that for many of these patients it was too late for treatment. Given the spatial component of the used analytical data model, it is now possible to characterize where theses patients lived and the regions that are more affected by this disease.

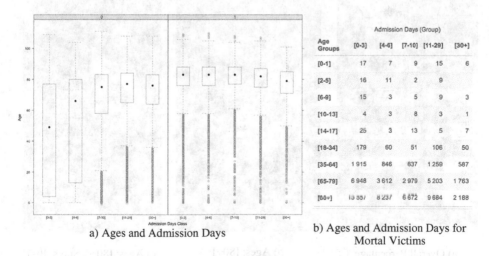

Age Groups	Admission Days (Group)				
	[0-3]	[4-6]	[7-10]	[11-29]	[30+]
[0-1]	17	7	9	15	6
[2-5]	16	11	2	9	
[6-9]	15	3	5	9	3
[10-13]	4	3	8	3	1
[14-17]	25	3	13	5	7
[18-34]	179	60	51	106	50
[35-64]	1 915	846	637	1 259	587
[65-79]	6 948	3 612	2 979	5 203	1 763
[80+]	13 557	8 237	6 872	9 684	2 188

a) Ages and Admission Days

b) Ages and Admission Days for Mortal Victims

Fig. 7. Relation between ages and number of days in the healthcare facility

Before proceeding with the data analytics study, and for a technological characterization of the used tools, it is worth mentioning that all the dashboards presented in
the following section were implemented using Tableau [20], while the graphs presented
in this section were implemented using Tableau or R [19].

5   Geo-Spatial Characterization of Pneumonia Victims

Given the context of the previous section, the number of fatalities, its increase all over
the years, and the fact that this disease seriously affects particular groups of people, this
section provides a geo-spatial characterization of these victims, trying to understand
this phenomena, knowledge that is essential for the appropriate definition of actions to
fight it. As shown in Fig. 8(a), with the overall percentage of victims attending to the
number of cases, the *Beja* district stands out with an average of 25.43 % of victims. In
general, the South and the interior part of the country are more affected by this disease.
If we restrict the data to those individuals with 80 or more years old (Fig. 8(b)), the
difference between North and South is even more noticeable, but now with the district
of *Setúbal* being more affected, with an average fatality rate of 39.35 %. If we continue
filtering data to consider now those victims with 80 or more years old and with very
short stays in the hospital ([0–3]), we can see that the percentage increases in all cases,
with an overall percentage of victims that is very high, reaching almost 90 % in
districts like *Beja* (89.27 %) or *Guarda* (84.01 %).

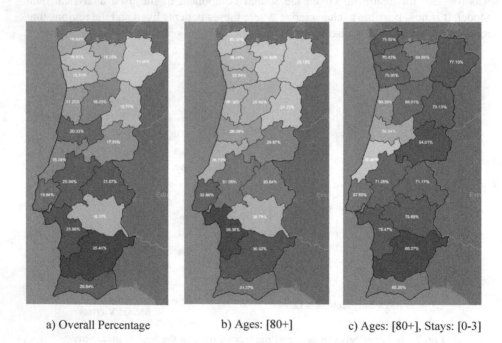

a) Overall Percentage   b) Ages: [80+]   c) Ages: [80+], Stays: [0-3]

Fig. 8. Percentage of mortal victims

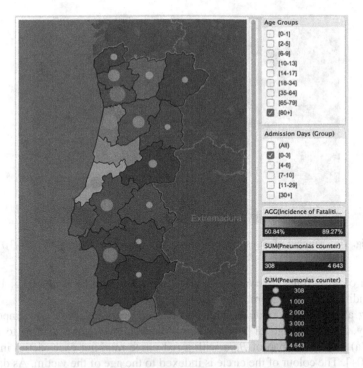

Fig. 9. Number of cases and percentage of victims ([80+], [0–3])

It is also important to stress that this behaviour is not only associated to these individuals, 80 or more years, as for the age class of [65–79], although with a smaller incidence, *Beja* presents, for example, a percentage of victims of 70.67 %. This is even more relevant if we consider that, for these regions, usually few cases of pneumonia are verified, although it seems that more severe. Considering the age class of 80 or more years old, the more affected one, Fig. 9 shows a dashboard applying a filter to this age class ([80+]), and to the shorter stays ([0–3]), and, as can be seen, more cases of pneumonia are verified in the metropolitan areas of *Lisboa* and *Porto*, but with a percentage of victims of 67.65 % for 4 643/3 141 cases of pneumonia/victims and 70.81 % for 2 364/1 675 cases of pneumonia/victims, respectively, contrasting with *Beja* and its 89.27 % for 317/283 cases of pneumonia/victims.

Looking to the particular case of *Beja*, it is now needed to drill-down and see what is the scenario inside the district. For that, the analysis of the several municipalities and parishes is useful, obtaining a higher detail in the geo-spatial characterization.

Figure 10 depicts the indicators under analysis for the municipality of *Beja* and an interesting pattern emerges. Six of the municipalities present 100 % of victims ([80+] for ages and [0–3] for stays) and all are located in the interior of the district. In this figure, the percentage of incidence of victims ranges from 73.33 % to 100 %, while the number of cases by municipality ranges from 2 to 75.

The analysis of this percentage, district by district, allowed the verification that different districts present different geo-spatial incidences, either with higher mortality to

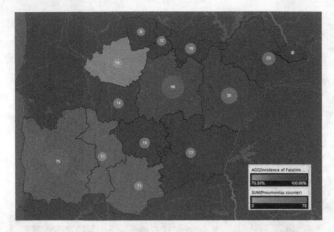

Fig. 10. Number of cases and percentage of victims for *Beja* ([80+], [0–3])

the interior of the country, like *Beja* (Fig. 10), to the littoral, like *Braga* (Fig. 11(a)), or with an undifferentiated pattern, like *Lisboa* (Fig. 11(b)).

Having all regions individuals with 80 or more years old, it is now important to verify why the percentage of victims is so different from one district to another. Figure 12(a) presents a map of *Beja* with a red circle marking each victim in the age class of [80+]. The colour of the circle is indexed to the age of the victim. As darker the circle, as older the victim, ranging ages from 80 to 101 years old. In this case, it seems that the municipalities with higher rates of mortality are the ones with eldest people, although no strong correlation was found between these two metrics. Figure 12(b) presents the values of the median and average for age in each municipality of *Beja* and the average value for the percentage of mortality. As can be seen, the difference between genders is relevant, being male in general affected sooner that female. This trend was verified in all the 18 districts of continental Portugal.

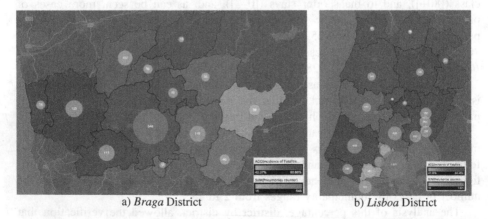

a) *Braga* District   b) *Lisboa* District

Fig. 11. Number of cases and percentage of victims for other districts ([80+], [0–3])

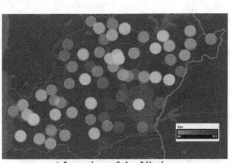

	Gender					
	F			M		
Municipality	Median Age	Avg. Age	Incidence of Fatalities	Median Age	Avg. Age	Incidence of Fatalities
Aljustrel	86.00	87.67	100.00%	83.00	83.63	87.50%
Almodôvar	83.50	84.67	83.33%	89.00	86.60	80.00%
Alvito	91.00	90.33	100.00%	86.00	85.83	100.00%
Barrancos	85.50	85.50	100.00%			
Beja	90.00	89.03	82.76%	87.00	86.46	91.89%
Castro Verde	87.00	88.88	100.00%	87.00	87.00	100.00%
Cuba	85.50	86.63	100.00%	85.50	87.00	100.00%
Ferreira do Alentejo	87.00	87.33	77.78%	86.00	86.33	66.67%
Mértola	87.00	88.00	100.00%	86.00	87.00	100.00%
Moura	85.00	86.57	92.86%	84.00	85.00	100.00%
Odemira	86.00	88.34	72.41%	84.50	86.07	89.13%
Ourique	85.00	85.00	80.00%	84.00	84.67	83.33%
Serpa	86.00	87.00	81.25%	85.00	85.05	100.00%
Vidigueira	86.00	85.67	100.00%	86.00	87.57	100.00%
Grand Total	86.00	87.59	86.00%	85.00	86.07	92.22%

a) Location of the Victims   b) Age and Percentage of Mortality

Fig. 12. Spatial distribution of the victims in *Beja* ([80+], [0–3]) (Color figure online)

In general, and taking as an example the three districts more detailed until now, we can look into the number of readmissions each patient had (Fig. 13). Considering all patients, all ages and limiting the analysis to the shorter stays ([0–3] days), in general *Beja* presents fewer readmissions for each patient and, as already seen, higher mortality, a phenomenon that is, for this district, also verified in younger patients. In the case of no readmission, *Braga* and *Lisboa* present a crescent trend pattern related with age, which is associated to the number of pneumonia cases. In the case of 1 or more readmissions, they are mostly verified after the sixties for *Beja*, after the forties for *Braga*, and after the twenties for *Lisboa*. Figure 13 limits the visualization to a maximum of two readmissions, although in some cases more readmissions were verified. In this figure, colours are associated with the defined age groups.

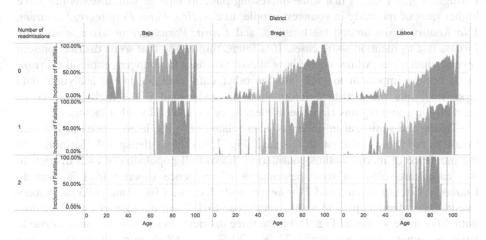

Fig. 13. Number of readmissions for shorter stays ([0–3] days)

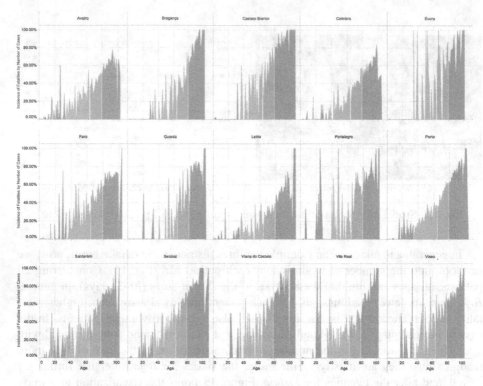

Fig. 14. Overall percentage of mortality for shorter stays ([0–3] days)

In an overall characterization of the several districts, the other 15 of continental Portugal, Fig. 14 shows that some interesting patterns emerge with districts that have higher rates of mortality in youngest people, like *Aveiro, Faro, Portalegre, Santarém, Vila Real* or *Viseu* around the twenties, and *Évora, Portalegre* or *Viseu* around the forties, just to mention some cases. It is interesting to see that some districts present several similarities, while others show almost no cases in younger people like *Évora*.

It is now important to look into the other data available in the analytical data repository, like the statistical information, to understand if the high incidence of mortality in some regions cloud be influenced or explained by other factors.

Taking the statistical information, data related with the latest census in Portugal (made in 2011) was selected. In this case, Fig. 15(a) shows the spatial distribution of the incidence of mortal victims considering the overall population of each district. In this case, three districts have percentages of incidence superior to 1 %, namely *Coimbra, Castelo Branco* and *Portalegre*, with 1.10 %, 1.04 % and 1.03 %, respectively. Other districts present values very close to 1 %. In the case of mortal victims with 80 or more years old, Fig. 15(b), the three districts already pointed out continue to have the higher values, now with 0.72 %, 0.70 % and 0.68 %, respectively. Only when the available information is filtered, considering the shorter stays in the hospital, Fig. 15(c), *Castelo Branco* presents the highest percentage of victims attending to the

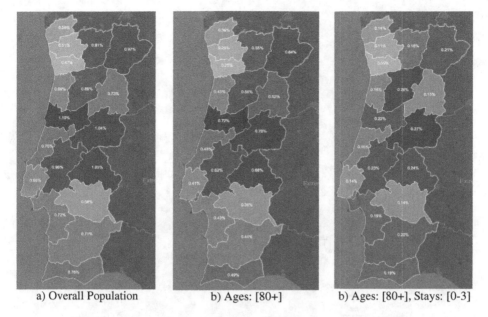

a) Overall Population   b) Ages: [80+]   b) Ages: [80+], Stays: [0-3]

Fig. 15. Percentage of mortal victims regarding the overall population

population of that district, namely 0.27 %, being followed by *Viseu* and *Portalegre*, with 0.26 % and 0.24 %, respectively.

In Fig. 15(a), one district called our attention due to its dissimilarity with the others also located at the littoral of the country. *Coimbra* presents the highest percentage of mortal victims attending to the global population in that district. Along the years under analysis, and as already mentioned, the overall increase in terms of the number of mortal victims was more than 65 %, and *Coimbra*, as can be see in Fig. 16, follows this average trend, with 66.15 %, having an overall incidence of mortality of 19.52 %. *Beja* is again in the spot not only because this district has the highest overall incidence of mortality, 35.78 %, but also because the variation of the number of victims was, from 2002 to 2011, of 143.94 %. *Castelo Branco* presents the highest variation with 167.68 % being followed by *Leiria* with 147.01 %.

Giving this context of overall variation of the incidence of mortality, it is now relevant to verify the evolution of the number of mortal victims along the years (Fig. 17). In Fig. 17(a), the variation of mortal victims considering the different age groups and the several years shows that, in younger patients, the variation is usually higher although few cases are verified. In these cases, some outliers show variations occasionally higher than 100 %, either positive or negative (those cases were filtered from the image for the sake of clarity). The variation of cases for the several years tends to avoid huge variations with age, being the number of victims a more constant number considering the number of pneumonia cases. However, along the years, the number of victims has increased considerably in the age class of 80 or more years old, as can be seen in Fig. 17(b). In global terms, the incidence of victims is around 30 % for this age class, less than 20 % for [65–79], less than 10 % for [25–64], and so on.

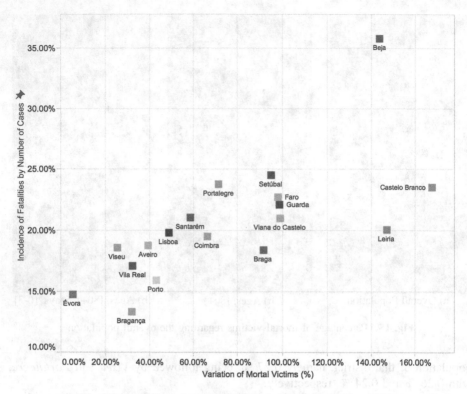

Fig. 16. Variation of the number of victims from 2002 to 2011

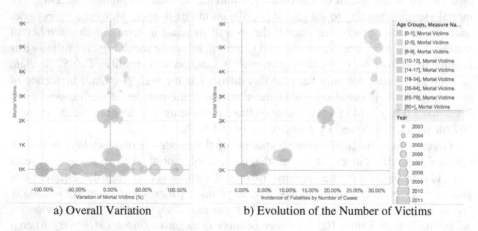

a) Overall Variation   b) Evolution of the Number of Victims

Fig. 17. Variation of the number of mortal victims along the years

For the three districts with the highest variations in the decade under analysis, *Beja*, *Castelo Branco* and *Leiria*, Fig. 18(a) shows how these districts behaved along the years. In the case of *Leiria*, this district presents an increase in terms of the age classes [65–79] and [80+] that is very impressive. Although with a significant increase in the

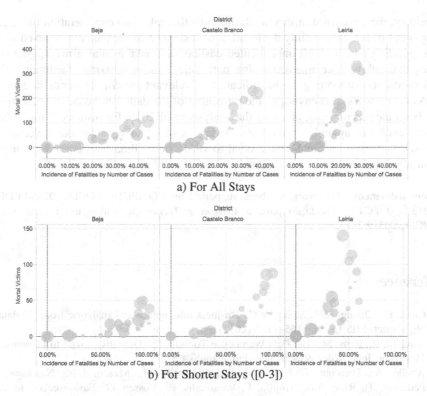

a) For All Stays

b) For Shorter Stays ([0-3])

Fig. 18. Relevant variations of the number of mortal victims along the years

number of victims, the percentage of mortality is lower than the verified in *Beja* or *Castelo Branco*, even when the number of days of admissions, to consider the shorter stays, is filtered (Fig. 18(b)). Each one of these districts present a characteristic trend in the evolution of the disease and its consequences. Even in a small country like Portugal, the differences between districts is so high that justify a deepest analysis and the identification of the potentiating factors.

Given the presented analyses, it is possible to see that data, when properly stored in an analytical repository, can be analysed in an interactive way, combining different perspectives and applying different filters to data. The goal is to gain a deeper understanding of the phenomena under analysis, in order to support the decision making process. In this case, the purpose was to spatially characterize a disease that provokes so many deaths. The knowledge obtained should allow decision makers to define appropriate measures to fight this disease.

6　Conclusions and Future Work

This paper presented an overall geo-spatial characterization of pneumonia incidence in continental Portugal, taking into consideration, mostly, the mortal victims caused by this disease. Data from 10 years counting 369 160 records, available in an analytical repository, were analysed in specific dashboards that take into consideration the spatial component of the data, mainly the location of residence of the patients, indexed to the corresponding parishes. All implemented dashboards make available maps, graphs or tables that allow user interaction for data selection or filtering, facilitating data exploration and supporting the identification of relevant patterns or trends in data.

As future work, it is envisaged the refreshing of the data warehouse in order to add data from the recent years, allowing the analysis until 2015, for instance. Moreover, it is envisaged the upgrade of the data model, in order to consider other vectors of analysis, like environmental data, crucial to verify how climacteric conditions or pollution affect the course of this disease.

Acknowledgement. This work has been supported by COMPETE: POCI-01-0145-FEDER-007043 and FCT – *Fundação para a Ciência e Tecnologia* within the Project Scope: UID/CEC/00319/2013.

References

1. Chen, H., Chiang, R.H., Storey, V.C.: Business intelligence and analytics: from big data to big impact. MIS Q. **36**, 1165–1188 (2012)
2. Kimball, R., Ross, M.: The Data Warehouse Toolkit: The Definitive Guide to Dimensional Modeling. John Wiley & Sons Inc., Indianapolis (2013)
3. Abelló, A., Darmont, J., Etcheverry, L., Golfarelli, M., Mazón, J.-N., Naumann, F., Pedersen, T., Rizzi, S.B., Trujillo, J., Vassiliadis, P., Vossen, G.: Fusion cubes: towards self-service business intelligence. Int. J. Data Warehouse. Mining **9**, 66–88 (2013)
4. Han, J., Kamber, M., Pei, J.: Data Mining: Concept and Techniques. Morgan Kaufmann Publishers, San Francisco (2012)
5. Viswanathan, G., Schneider, M.: On the requirements for user-centric spatial data warehousing and SOLAP. In: Xu, J., Yu, G., Zhou, S., Unland, R. (eds.) DASFAA Workshops 2011. LNCS, vol. 6637, pp. 144–155. Springer, Heidelberg (2011)
6. Rivest, S., Bédard, Y., Proulx, M.-J., Nadeau, M., Hubert, F., Pastor, J.: SOLAP technology: merging business intelligence with geospatial technology for interactive spatio-temporal exploration and analysis of data. ISPRS J. Photogrammetry Remote Sensing **60**, 17–33 (2005)
7. Santos, M.Y., Leite, V., Carvalheira, A., de Araújo, A.T., Cruz, J.: Characterization of pneumonia incidence supported by a business intelligence system. In: Ortuño, F., Rojas, I. (eds.) IWBBIO 2015, Part I. LNCS, vol. 9043, pp. 30–41. Springer, Heidelberg (2015)
8. Eurostat: Respiratory diseases statistics, June 2016. http://ec.europa.eu/eurostat/statistics-explained/index.php/Respiratory_diseases_statistics
9. WHO: World Health Organization. "The top 10 causes of death." 27 May 2015 (2015). http://who.int/mediacentre/factsheets/fs310/en/
10. Sufahani, S.F., Razali, S.N.A.M., Mormin, M.F., Khamis, A.: An analysis of the prevalence of pneumonia for children under 12 year old in Tawau general hospital, Malaysia. In: Proceedings of the International Seminar on the Application of Science & Mathematics, Kuala Lumpur (2011)
11. Oroszi, F., Ruhland, J.: An early warning system for hospital acquired pneumonia. In: Proceedings of the 18th European Conference on Information Systems (2010)

12. Trillo-Alvarez, C., Cartin-Ceba, R., Kor, D.J., Kojicic, M., Kashyap, R., Thakur, S., Thakur, L., Herasevich, V., Malinchoc, M., Gajic, O.: Acute lung injury prediction score: derivation and validation in a population-based sample. Eur. Respir. J. **37**, 604–609 (2011)
13. Wu, C., Rosenfeld, R., Clermont, G.: Using data-driven rules to predict mortality in severe community acquired pneumonia. PLoS ONE **9**, e89053 (2014)
14. Peng, R.: Exploratory data analysis with R (2015). http://Lulu.com
15. Tufte, E.R.: Beautiful Evidence, 1st edn. Graphics Press, Cheshire (2006)
16. Tufte, E.R., Graves-Morris, P.R.: The Visual Display of Quantitative Information. Graphics press, Cheshire (1983)
17. INE: Portugal census (2011). http://censos.ine.pt
18. Santos, M.Y., Carvalheira, A., de Araujo, A.T.: A data-driven analytics approach in the study of pneumonia's fatalities. In: IEEE International Conference on Data Science and Advanced Analytics (DSAA), 36678 2015, pp. 1–10. IEEE (2015)
19. R-project: the R project for statistical computing (2016). https://www.r-project.org
20. Tableau (2016). http://www.tableau.com

Ontology-Guided Principal Component Analysis: Reaching the Limits of the Doctor-in-the-Loop

Sandra Wartner[1], Dominic Girardi[1]([✉]), Manuela Wiesinger-Widi[1],
Johannes Trenkler[2], Raimund Kleiser[2], and Andreas Holzinger[3]

[1] Research Unit Medical Informatics at RISC Software GmbH,
Johannes Kepler University, Hagenberg and Linz, Austria
{sandra.wartner,dominic.girardi,manuela.wiesinger-widi}@risc-software.at
[2] Institute of Neuroradiology
at Neuromed Campus of the Kepler University Klinikum, Linz, Austria
[3] Research Unit, HCI-KDD, Institute for Medical Informatics,
Statistics and Documentation, Medical University Graz, Graz, Austria

Abstract. Biomedical research requires deep domain expertise to perform analyses of complex data sets, assisted by mathematical expertise provided by data scientists who design and develop sophisticated methods and tools. Such methods and tools not only require preprocessing of the data, but most of all a meaningful input selection. Usually, data scientists do not have sufficient background knowledge about the origin of the data and the biomedical problems to be solved, consequently a doctor-in-the-loop can be of great help here. In this paper we revise the viability of integrating an analysis guided visualization component in an ontology-guided data infrastructure, exemplified by the principal component analysis. We evaluated this approach by examining the potential for intelligent support of medical experts on the case of cerebral aneurysms research.

Keywords: Principal component analysis · Ontology · Data mining · PCA · Data warehousing · Doctor-in-the-loop

1 Introduction

Medicine is constantly turning into a data intensive science and the quantity of available health data is enormously increasing - far beyond what a medical doctor can handle [4]. Within such large amounts of data, relevant *structural* and/or *temporal* patterns ("knowledge") are often hidden and not accessible to the medical doctors [14].

However, the real problem is not only in the large quantities of data (colloquially called: "big data"), but in "complex data". Medical doctors today are confronted with complex data sets in arbitrarily high dimensions, mostly heterogeneous, semi-structured, weakly-structured and often noisy [15] and of poor data quality. The handling and processing of this data is known to be a major technical obstacle for (bio-)medical research projects [2]. However, it is not only

© Springer International Publishing Switzerland 2016
M.E. Renda et al. (Eds.): ITBAM 2016, LNCS 9832, pp. 22–33, 2016.
DOI: 10.1007/978-3-319-43949-5_2

the data handling that contains major obstacles, also the application of advanced data analysis and visualization methods is often only understandable for data scientists. This situation will become even more dramatic in the future due to the ongoing trend towards personalized medicine with the goal of tailoring the treatment to the individual patient [12].

Interestingly, there is evidence that human experts sometimes still outperform sophisticated algorithms, e.g., in the instinctive, often almost instantaneous interpretation of complex patterns. A good example is diagnostic radiologic imaging, where a promising approach is to fill the semantic gap by integrating the physicians high-level expert knowledge into the retrieval process by acquiring his/her relevance judgments regarding a set of initial retrieval results [1].

Consequently, the integration of the knowledge of a domain expert may sometimes greatly enhance the knowledge discovery process pipeline. The combination of both human intelligence and machine intelligence, by putting a "human-in-the-loop" would enable what neither a human nor a computer could do on their own. This human-in-the-loop can be beneficial in solving computationally hard problems, where human expertise can help to reduce an exponential search space through heuristic selection of samples, and what would otherwise be an NP-hard problem, reduces greatly in complexity through the input and the assistance of a medical doctor into the analytics process [13]. This approach is supported by a synergistic combination of methodologies of two areas that offer ideal conditions towards unraveling such problems: Human-Computer Interaction (HCI) and Knowledge Discovery/Data Mining (KDD), with the goal of supporting human intelligence with machine intelligence to discover novel, previously unknown insights into data (HCI-KDD approach [11]).

From the theory of human problem solving it is known that, for example, medical doctors can often make diagnoses with great reliability – but without being able to explain their rules explicitly [6]. Here this approach could help to equip algorithms with such "instinctive" knowledge. The importance of this approach becomes clearly apparent when the use of automated solutions due to the incompleteness of ontologies is difficult [3].

The immediate integration of the domain expert into data exploration has already proved to be very effective, for example in knowledge discovery [9], or in subspace clustering [17], compelling the domain expert to face the major challenge of detecting mutual influences of variables. Having already an idea of those dependencies, the domain expert's goal, here: the medical doctor, is to confirm his suspicions; contrary to the data scientist, who has hardly any domain knowledge and therefore no insight in reasonable input for specific tools. Frequently, for many domain experts it is even not possible to have access to worthwhile, already long-time existing data analysis tools, including, e.g., the Principal Component Analysis (PCA), due to a lack of mathematical knowledge, on the one side, and missing computational knowledge, on the other side. Consequently, the role of the domain expert turns from a passive external supervisor – or customer – to an active actor of the process, which is necessary due to the enormous complexity of the medical research domain [5].

A survey from 2012 among hospitals from Germany, Switzerland, South Africa, Lithuania, and Albania [23] showed that only 29 % of the medical personnel of responders were familiar with practical applications of data mining. Although this survey might not be representative globally, it clearly shows the trend that medical research is still widely based on standard statistical methods. One reason for the rather low acceptance rate of data mining tools is the relatively high technical obstacle that often needs to be taken in order to apply complex algorithms combined with the limited knowledge about the algorithms themselves and their output. Especially in the field of exploratory data analysis deep domain knowledge of the human expert is a crucial success factor.

In order to address this issue, we developed a data infrastructure for scientific research that actively supports the domain expert in tasks that usually require IT knowledge or support, such as: structured data acquisition and integration, querying data sets of interest by non-trivial search conditions, data aggregation, feature generation for subsequent data analysis, data preprocessing, and the application of advanced data visualization methods. It is based upon a generic meta data model and is able to store the current domain ontology (formal description of the actual research domain) as well as the corresponding research data. The whole infrastructure is implemented at a higher level of abstraction and derives its manifestation and behavior from the actual domain ontology at run-time. Just by modeling the domain ontology, the whole system, including electronic data interfaces, web portal, search forms, data tables, etc., is customized for the actual research project. The central domain ontology can be changed and adapted at any time, whereas the system prevents changes that would cause data loss or inconsistencies. In this context, medical experts are offered assistance in their research purposes.

In many cases, these domain experts are unfamiliar with the variety of mathematical methods and tools which greatly simplify data exploration. In order to overcome impediments concerning mathematical expertise or the selection and application of suitable methods, we propose ontology-guided implementations for domain-expert-driven data exploration. One of those major methods is Principal Component Analysis (hereinafter referred to as PCA, see Sect. 2), representing a powerful method for dimensionality reduction.

In order to assist domain experts data preprocessing is automated as far as possible using the user-defined domain-ontology to overcome technical obstacles already in advance. Thus, the domain expert is capable of performing the fundamental analysis on his own. By selecting data of interest and starting the calculations, PCA is run in the background and results in an inbuilt visualization for more convenient access of data information.

2   Principal Component Analysis

Principal Component Analysis (PCA) is a method for reducing the dimension of a data set such that the new set contains most of the information of the original set and can be interpreted more easily. PCA was first described by Pearson [25] and since then has been reinvented in different fields such as Economic Sciences [16], Psychology [28,29], and Chemistry [20,27] under different names like Factor Analysis or Singular Value Decomposition. Also in other fields, including Geo Sciences and Social Sciences, PCA is an established method. For a good introduction to PCA see for example [18,26].

In the following paragraph we sketch the main idea of PCA. We are given a set of observations of m variables. PCA then computes the direction of maximal variance in these data in m-dimensional space. This direction forms a new variable (a linear combination of the original variables), the first principal component. This process is repeated with the remaining variance of the data until a specified number of principal components is reached or a specified percentage of the original variance is covered (explained variance of the system). Every succeeding principal component is orthogonal to the preceding ones and adds to the explained variance of the new system. There cannot be more principal components than original variables and if their number is equal then the explained variance is 100 %. Mathematically, PCA is a solution to the eigenvalue problem of the covariance resp. correlation matrix of the original variables where the eigenvectors form the principal components and the eigenvalues indicate the importance of the components (the higher the eigenvalue, the higher the explained variance of the component).

Of interest in interpreting the results of a PCA are the scores (projection of the original data points into the new vector space), loadings (eigenvectors multiplied by the square root of the corresponding eigenvalues, i.e., the loadings also include variance along the principal components), residuals and their respective plots. The score plot depicts the scores with respect to two selected principal components that form the axes of the plot. It is used to detect outliers and patterns in the data. The loadings plot depicts the original variables with respect to two selected principal components that form the axes of the plot. It is used to examine correlations between the original variables and to examine the extent to which the variables contribute to the different principal components. The biplot displays both scores and loadings simultaneously and allows to investigate the influence of the variables on the individual data points or groups of data points, respectively.

First ideas of introducing PCA in the medical field came up in the early 70 s, gradually increasing. From 2006 onwards, the annual increase of research results is still growing very fast, comprising already about 670 scientific results in 2015 on NCBI [22]. Currently, PCA establishes a satisfying solution in various medical sub domains for different purposes. The application field ranges from image processing, like image compression or recognition [21], to data representation, for facilitating analysis.

3   Ontology-Guided PCA

In this section, we briefly review the main integration actions of the PCA method (see Sect. 2) into the data infrastructure. Above all the term ontology has to be clarified as there is a degree of uncertainty around the terminology, whereby for computer scientists an ontology is described as formal descriptions, properties and relationships between objects in the world [30].

3.1   Ontology-Guided Clinical Research Infrastructure

The theoretic base for the already mentioned ontology-based research infrastructure is a revision and adaption of the established process models for knowledge discovery. In the commonly known definitions of this process (see [19] for a good overview) the domain-expert is seen in a supervising, consulting and customer role. A person who is outside the process and assists in crucial aspects with domain knowledge and receives the results. All the other steps of the process are performed by so called data analysts, who are supported by the domain-experts in understanding the for the current research project relevant aspects of the research domain and interpreting the results. We revised these process models and proposed a new, domain-expert-centric process model for medical knowledge discovery [8]. Based upon this process model we developed a generic research infrastructure, which supports the domain expert throughout the whole process — from data model, over data acquisition and - integration, data processing, and quality-management to data exploration. The research infrastructure is domain independent and derives its current appearance and behavior from the user-defined domain ontology at run-time. The researcher is able to define what data structure he or she needs to answer the research questions. This definition — the domain ontology — then builds the base for the whole system. From a user's point of view, the infrastructure consists of three main modules:

1. The Management Tool: This Java rich-client application allows the user to defined and maintain the domain ontology. Furthermore, the whole data set of the system can be searched, filter, processed and analyzed in this application. The here presented work is integrated into this application.
2. The Data Integration Module: This module is a plug-in to an established open-source ETL (Extract-Transform-Load) Suite. It allows to access structured data from almost arbitrary sources and to properly integrate this data into the research infrastructure.
3. The Web Interface: If data needs to be acquired manually, the web interface offers domain-derived forms to view, enter, process the data via a web browsers. In the clinical context this is often necessary when information from semi- or unstructured documents (e.g. doctor's letters, care instructions, etc.) needs to be stored in a structured way.

For more detailed information on this infrastructure the reader is kindly referred to [7].

3.2   Background Processes

The execution of PCA requires structured processing of data. In our data infrastructure all of those preparatory steps are based on ontological meta-information and are automatically performed in the background. In this case the domain expert neither has to be concerned about data types, data transformation, starting the corresponding algorithm nor about collecting and depicting results. Therefore, solely a few steps remain, explicitly data selection and parameter setting, in order to start PCA. After variable selection out of a (sub) set of data and adjusting parameter configuration PCA performs the projection into lower dimensional space. The result is visualized in interactive two dimensional charts (loading-, score- and biplot), offering manipulation of axes and therefore displaying different combinations of principal components. Backtracking to the pristine records can establish a better idea of relationships when selecting the scores. In a few steps data is ready for analysis.

3.3   Implementation

For the actual implementation, we used the WEKA library [10] for performing PCA. Therefore some partial integration has been necessary in order to acquire mathematical PCA output for visualization purposes.

Step 1. First, an ontology-guided transformation of the data into the WEKA **data** structure (*weka.core.Instances*) and converting our variables to WEKA conform **attributes** (*weka.core.Attribute*) has to be performed. An **evaluator** (*weka.attributeSelection.PrincipalComponents*, defining the evaluation method) has to be configured by setting the variance covered by the principal component as well as whether the correlation or the covariance matrix has to be used. All of those transformations are performed in an ontology-guided, hidden behavior from the researcher's perspective.

Step 2. The key component of the WEKA PCA is represented and performed by the **feature selection** (*weka.attributeSelection.AttributeSelection*), hence a **ranker** (*weka.attributeSelection.Ranker*, defining the search method) as well as the evaluator configured in step 1 have to be assigned. The specified ranker's task is to supervise whether the defined threshold (explained variance) or a specified number of components is reached, thus PCA has finished.

Step 3. After completion of the embedded WEKA PCA process, an internal PCA result class is prepared, carrying the most essential output including eigenvectors and eigenvalues, scores and loadings as well as the number of principal components and proposed features. Accordingly, the generated output is subject to back transformation in the prescribed ontology and is processed for being displayed in a scatter chart related visualization.

Step 4. In order to determine the quality of the result some key figures have to be determined. Therefore we take into account linearity of the input data, the size of the data set, the variance covered as well as tests on normal distribution. The outcome of the quality test is displayed within the visualization, supporting the researcher in evaluating the significance of the outcome. Since PCA is vulnerable to outliers, outlier detection is provided in the visualization, making it possible for the user to exclude these points and re-initiate a PCA. In particular, the rationale for various quality outcomes is the quality of the input data.

4   Results

We evaluated the viability of this approach to perform PCA within an ontology-guided data infrastructure for scientific research purposes on a data set of 1237 records, representing a cerebral aneurysm each. This vulnerability of a blood vessel is described as the dilation, ballooning-out, or bulging of part of the wall of an artery in the brain [24]. Those samples were taken from patients, registered by the Institute of Neuroradiology at Neuromed Campus of the Kepler University Klinikum. The aim of this collaboration was to collect and analyze the medical outcome data of their patients, who have cerebral aneurysms. The main research subject of the database is the clinical and morphological follow up of patients with cerebral aneurysms, which were treated either with an endovascular procedure, surgically or conservative [9].

We attempt to show the feasibility of the ontology-based research, done by the domain expert without assistance of a data scientist. In this context the domain experts are from the field of neurosurgery and neuro-radiology. The following parameters of the aneurysm were taken into account: the age of patient at diagnosis, number of aneurysms in total for this patient, the number of recorded clinical events (complications), the number of surgical treatments, the number of endovascular treatments, as well as the width, depth, height, neck and size of an aneurysm. It was not aim of this evaluation to discover new medical knowledge, but rather to verify the method by being able to demonstrate already known knowledge about the data.

The result in form of the loadings plot is shown in Fig. 1. It shows the first two principal components (PC) with the highest percentage of explained variance. The first PC is displayed on the x-axis and the second PC is shown on the y-axis. It is apparent from this plot that there is a strong relationship between the *Width*, *Depth* and *Height* of aneurysms, as they are located close to one another. From a medical point of view, this is obvious, since aneurysms are spheric in most of the cases. Another variable is in the surrounding of this variable cluster, namely the variable *Neck*. The neck describes the diameter of the opening of the aneurysm to the supplying blood vessel. Here again, a correlation is indicated by the nature of the matter. The bigger the aneurysm is in all its dimensions, the bigger the neck tends to be. On the other principal component, the opposing position of the number of endovascular treatments on the one hand and surgical treatments on the other side is appealing. This is given due to the fact, that most aneurysms are either treated the one or the other way. The position of the variable *Age of Patient at Diagnosis* very close to the center of the visualization indicates that there are hardly any correlations between this variable and the others and this variable has no influence on the shape of the data cloud.

While the previous observations were easily explained with already known facts, the opposing position of the width-depth-height-cluster on the left-hand-side of the first PC and the variable *Total Number of Aneurysms for Patient* on the right-hand-sind struck the attention of the medical researchers. Preceding visualization with other methods already indicated a (weak) reverse correlation between the total number of aneurysms a patient suffers from and the size of

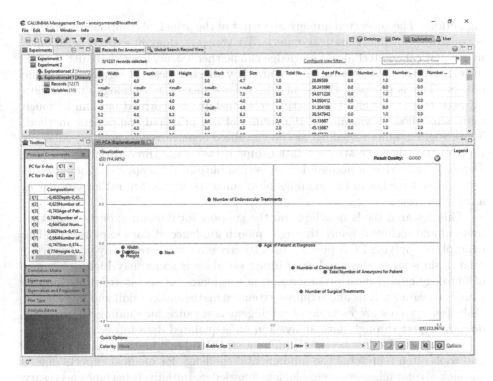

Fig. 1. A two dimensional loadings plot of the aneurysm data set, embedded in an ontology-guided data infrastructure. The x-axis represents the first principal component, expressing 33.96 % of the total variance. The second principal component conveys 14.98 %, depicted by the y-axis.

these aneurysms. All methods, including this PCA-run, showed evidence, that patients with numerous aneurysms tend to have smaller ones. This phenomenon will now be investigated. This is a very good example for what the ontology-guided approach for the doctor-in-the-loop is able to do. It allows the researching domain experts to explore their complex data and generate new hypotheses for subsequent research.

The automatic calculation of the relevant key figures indicate that this PCA-run yielded an acceptable and meaningful result. The covered percentage of the variance is acceptable and colored in green. The result of this key number calculation and interpretation is visualized in the upper right corner of the visualization and giving an indication to the researcher how reliable this output is.

5 Discussion and Conclusion

For considerably complex mathematical methods results cannot be interpreted unambiguously at first glance, contrary to a simple bar chart or box plot. However, this is all the more important to provide guidance throughout data processing actions. The generated numerical output of the principal components method is conclusive for mathematical experts. By contrast, domain experts with basic mathematical and technical knowledge can neither see any immediate use regarding eigenvectors and eigenvalues, nor are they able to assess the significance of the output. This is precisely the point where assistance of an ontology-guided data infrastructure takes effect. Detecting relationships and correlations can be much more simplified by visualizing the results of the principal component method. Thus, Fig. 1 is quite revealing, as the visualized eigenvectors (loadings) substantially better illustrate strong relationships between the three variables (*Width*, *Depth*, *Height*) than a non-guided numerical output. As suspected, they are situated close together in the loadings plot, since aneurysms are rather circular in almost all cases.

This research sheds new light on the support for domain experts in mathematical and technical issues through smooth guidance of data exploration. The example of applying PCA pursues the objective to reveal previously unsuspected relationships when the number of input variables is supremely high. This way of ontology-guided data preprocessing is considered as an intermediate step in (medical) data analysis and requires extensive mathematical skill and knowledge. Only few systems are capable of intelligent assistance for guiding the medical domain expert through data analysis in an acquainted data infrastructure.

Quick and effortless access to different statistical and mathematical methods and tools often represents the fundamental challenge for medical experts due to the lack of comprehensive technological knowledge. Initially, it becomes necessary to give domain experts an understanding of the variety of available methods. Even if an appropriate method has been found, the major obstacle is the related realization, provided that the researcher is aware of mathematical science.

Not all results of a PCA-run are equally good and meaningful. It is very dangerous to use the PCA without further exploration of some statistical key numbers. The visualization tries to bring the result of the automatic calculation of these key features to the user. The percentage of the covered variance is colored, in a range from green (acceptable) over orange to red (unacceptable) (see Fig. 1). These two and the other key numbers are summarized in the info field *Result Quality* in the upper right corner of the visualization. There is of course no clear cut between the qualities of PCA results, but the sum of key numbers and their coloring provides guidance to the user to interpret the results. For very inexperienced users it provides a first barrier to use PCA results without any control of the key figures. This aspect clearly distinguishes the PCA from our preceding attempts (e.g., [9]) to make complex data mining and data visualization algorithms accessible to researching domain experts.

We realized that not all data mining and data visualization algorithms are meant to be used by non-data-scientists. We consequently try to push the technical barrier towards complex methods and algorithms in order to enable the biomedical domain experts to take advantage of them. Thus far, the results of these methods (non-linear mapping, parallel coordinates, etc.) were easy to interpret with limited danger to mis-interpretation. Here, in the case of PCA even very promising looking visualization might be completely worthless, and without checking the corresponding key-figures an interpretation is not possible. Only an intelligent research-platform designed for domain-expert-driven knowledge discovery can help by automatically calculating these key-figures and bringing them to a prominent position in the user interface.

6 Open Challenges and Future Work

Since the feasibility for PCA only subsists as numerical attributes are used, a small part of variables can be taken into consideration. Further work is required to establish the viability of extensive and automated ontology-based data preprocessing. Thus, especially in the medical domain, information is stored in categorical or boolean attributes. Extending the data infrastructure will lead to a PCA for categorical variables (multiple correspondence analysis, factor analysis for mixed data).

References

1. Akgul, C.B., Rubin, D.L., Napel, S., Beaulieu, C.F., Greenspan, H., Acar, B.: Content-based image retrieval in radiology: Current status and future directions. J. Digit. Imaging **24**(2), 208–222 (2011)
2. Anderson, N.R., Lee, E.S., Brockenbrough, J.S., Minie, M.E., Fuller, S., Brinkley, J., Tarczy-Hornoch, P.: Issues in biomedical research data management and analysis: needs and barriers. J. Am. Med. Inf. Assoc. **14**(4), 478–488 (2007)
3. Atzmüller, M., Baumeister, J., Puppe, F.: Introspective subgroup analysis for interactive knowledge refinement. In: Sutcliffe, G., Goebel, R. (eds.) FLAIRS Nineteenth International Florida Artificial Intelligence Research Society Conference, pp. 402–407. AAAI Press, Menlo Park (2006)
4. Buchan, I.E., Winn, J.M., Bishop, C.M.: A unified modeling approach to data-intensive healthcare. In: Hey, T., Tansley, S., Tolle, K. (eds.) The fourth paradigm: Data-Intensive Scientific Discovery, pp. 91–98. Microsoft Research, Redmond (2009)
5. Cios, K.J., William Moore, G.: Uniqueness of medical data mining. Artif. Intell. Med. **26**(1), 1–24 (2002)
6. Gigerenzer, G., Gaissmaier, W.: Heuristic decision making. Ann. Rev. Psychol. **62**, 451–482 (2011)
7. Girardi, D., Dirnberger, J., Giretzlehner, M.: An ontology-based clinical data warehouse for scientific research. Saf. Health **1**(1), 1–9 (2015)
8. Girardi, D., Kueng, J., Holzinger, A.: A domain-expert centered process model for knowledge discovery in medical research: putting the expert-in-the-loop. In: Guo, Y., Friston, K., Aldo, F., Hill, S., Peng, H. (eds.) BIH 2015. LNCS, vol. 9250, pp. 389–398. Springer, Heidelberg (2015)
9. Girardi, D., Küng, J., Kleiser, R., Sonnberger, M., Csillag, D., Trenkler, J., Holzinger, A.: Interactive knowledge discovery with the doctor-in-the-loop: a practical example of cerebral aneurysms research. Brain Inf., 1–11 (2016). (Online First Articles)

10. Hall, M., Frank, E., Holmes, G., Pfahringer, B., Reutemann, P., Witten, I.H.: The WEKA data mining software: an update. ACM SIGKDD Explor. Newsl. **11**(1), 10–18 (2009)
11. Holzinger, A.: Human-computer interaction and knowledge discovery (HCI-KDD): what is the benefit of bringing those two fields to work together? In: Cuzzocrea, A., Kittl, C., Simos, D.E., Weippl, E., Xu, L. (eds.) CD-ARES 2013. LNCS, vol. 8127, pp. 319–328. Springer, Heidelberg (2013)
12. Holzinger, A.: Trends in interactive knowledge discovery for personalized medicine: Cognitive science meets machine learning. IEEE Intell. Inf. Bull. **15**(1), 6–14 (2014)
13. Holzinger, A.: Interactive machine learning for health informatics: When do we need the human-in-the-loop? Springer Brain Inform. (BRIN) **3**, 1–13 (2016). http://dx.doi.org/10.1007/s40708-016-0042-6
14. Holzinger, A., Dehmer, M., Jurisica, I.: Knowledge discovery and interactive data mining in bioinformatics - state-of-the-art, future challenges and research directions. BMC Bioinform. **15**(S6), I1 (2014)
15. Holzinger, Andreas, Stocker, Christof, Dehmer, Matthias: Big complex biomedical data: towards a taxonomy of data. In: Obaidat, Mohammad S., Filipe, Joaquim (eds.) ICETE 2012. CCIS, vol. 455, pp. 3–18. Springer, Heidelberg (2014)
16. Hotelling, H.: Analysis of a complex of statistical variables into principal components. J. Educ. Psychol. **24**, 417–441 (1933)
17. Hund, M., Bhm, D., Sturm, W., Sedlmair, M., Schreck, T., Ullrich, T., Keim, D.A., Majnaric, L., Holzinger, A.: Visual analytics for concept exploration in subspaces of patient groups: Making sense of complex datasets with the doctor-in-the-loop. Brain Inf. **3**, 1–15 (2016)
18. Kessler, W.: Multivariate Datenanalyse: für die Pharma-Bio- und Prozessanalytik. Wiley-VCH Verlag GmbH & Co. KGaA, Weinheim (2007)
19. Kurgan, L.A., Musilek, P.: A survey of knowledge discovery and data mining process models. Knowl. Eng. Rev. **21**(01), 1–24 (2006)
20. Malinowski, E.: A thesis in two parts: application of factor analysis to chemical problems. Stevens Inst. Technol. **2**(1–2), 54–94 (1961)
21. Nandi, D., Ashour, A.S., Samanta, S., Chakraborty, S., Salem, M.A., Dey, N.: Principal component analysis in medical image processing: a study. Int. J. Image Min. **1**(1), 65–86 (2015)
22. National Center for Biotechnology Information: Mesh search for principalcomponent analysis and medicine (2016). http://www.ncbi.nlm.nih.gov/
23. Niakšu, O., Kurasova, O.: Data mining applications in healthcare: research vs practice. Databases Inf. Syst. BalticDB&IS **2012**, 58 (2012)
24. NIH: Cerebral Aneurysm Information Page (April 2010). http://www.ninds.nih.gov/disorders/cerebral_aneurysm/cerebral_aneurysm.htm
25. Pearson, K.: On lines and planes of closest fit to systems of points in space. Philos. Mag. **2**, 559–572 (1901)
26. Rencher, A.: Methods of Multivariate Analysis. Wiley Series in Probability and Statistics. Wiley, Chichester (2002)
27. Sharaf, M., Illman, D., Kowalski, B.: Chemometrics. Wiley, New York (1986)
28. Thurstone, L.: Multiple-factor Analysis: A Development and Expansion of The Vectors of Mind. The university of Chicago committee on publications in biology and medicine. University of Chicago Press, New York (1947)
29. Thurstone, L., Thurston, T.: Factorial Studies of Intelligence. Psychometrika monograph suplements. The University of Chicago press, Chicago (1941)

30. Wang, B.B., Mckay, R.I., Abbass, H.A., Barlow, M.: A comparative study for domain ontology guided feature extraction. In: Proceedings of the 26th Australasian Computer Science Conference vol. 16, pp. 69–78. Australian Computer Society, Inc. (2003)

Enhancing EHR Systems Interoperability by Big Data Techniques

Nunziato Cassavia, Mario Ciampi, Giuseppe De Pietro, and Elio Masciari[⊠]

ICAR-CNR, Rende, Italy
{nunziato.cassavia,mario.ciampi,
giuseppe.depietro,elio.masciari}@icar.cnr.it

Abstract. Information management in healthcare is nowadays experiencing a great revolution. After the impressive progress in digitizing medical data by private organizations, also the federal government and other public stakeholders have also started to make use of healthcare data for data analysis purposes in order to extract actionable knowledge. In this paper, we propose an architecture for supporting interoperability in healthcare systems by exploiting Big Data techniques. In particular, we describe a proposal based on big data techniques to implement a nationwide system able to improve EHR data access efficiency and reduce costs.

Keywords: Big data · Healthcare · Interoperability

1 Introduction

Nowadays, the availability of huge amounts of data from heterogeneous sources, exhibiting different schemes and formats and being generated at very high rates, led to the definition of new paradigms for their management – this problem is known with the name *Big Data* [3–6]. As a consequence of new perspective on data, many traditional approaches to data analysis result inadequate both for their limited effectiveness and for the inefficiency in the management of the huge amount of available information.

Therefore, it is necessary to rethink both the storage and access patterns to big data as well the design of new tools for data presentation and analysis. It is worth noticing that the problem of fast accessing relevant pieces of information arises in several scenarios such as world wide web search, e-commerce systems, mobile systems and social networks analysis to cite a few. Successful analyses for all the application contexts rely on the availability of effective and efficient tools for browsing data so that users may eventually extract new knowledge which they were not interested initially.

In this respect, also healthcare stakeholders have access to challenging knowledge integration and extraction problems. This information can be classified as big data, as they exhibit impressive volume and they are really heterogeneous and time varying. Moreover, pharmaceutical-industry experts are interested to analyze big data to obtain useful insights on their data. Although these efforts

© Springer International Publishing Switzerland 2016
M.E. Renda et al. (Eds.): ITBAM 2016, LNCS 9832, pp. 34–48, 2016.
DOI: 10.1007/978-3-319-43949-5_3

are still in their early stages, they could help for providing people better health-care quality and reducing costs. As an example, it is possible to analyze patient data for understanding what treatments are most effective for particular conditions, identifying patterns related to drug side effects or hospital readmissions, and gaining additional important knowledge [18].

Indeed, a relevant research issue is the design and implementation of a distributed platform for accessing heterogenous Healthcare Information Systems (HIS) so as to enlarge their coverage and allowing the availability of medical data at heterogeneous, and geographically-sparse, healthcare providers by enabling data sharing in a seamless manner. In this paper we propose an architecture that overcomes the available architectures for federating HIS, as current solutions suffer the limitation of only allowing a communication style according to a pull-based data delivery, i.e., the user requests a clinical document knowing its unique reference. On the contrary, we take advantage of big data architectural advantages for offering a reliable solution that can be used also by non IT experts. In particular, we exploit the well know MapReduce framework in order to offer advanced querying capabilities for medical data.

2 Background and Related Work

2.1 Basics on Tools Supporting the Management of Big Data

Nowadays, dealing with a big volume of data is a very difficult challenge, since traditional technologies, like RDBMS, are not suited for this purpose. Many open source technologies were developed in order to handle massive amounts of data. The majority of these technologies are based on the MapReduce programming model. This paradigm make it easier to implement solutions based on the use of distributed systems for executing data mining tasks.

The MapReduce paradigm is based on the following steps:

- *Map*: Each node executes the *map* function on its local data, creating a set of pairs ⟨*key, value*⟩, and stores the results in a temporary storage.
- *Shuffle*: Pairs ⟨*key, value*⟩ are redistributed among nodes, in such a way that all the pairs with the same key are assigned to the same node.
- *Reduce*: Each node processes its group of pairs, independently of other nodes.

It is worth noticing that since each mapping operation does not depend on the others, mapping operations can be parallely executed. In a similar way, also the reduce step can be performed by multiple nodes at the same time, if the reduction function is associative.

The most widespread implementation of the MapReduce programming model is Hadoop MapReduce, part of the Hadoop framework [23]. Although Hadoop is a really pervasive technology, it has its drawbacks, especially with clustering (or in general with machine learning) algorithms based on iterative operations. This is because Hadoop MapReduce stores the results of intermediate computations on disk. The overhead to launch each job, moreover, is very high. MapReduce is well suited for large distributed data processing where fast performance is not an issue. Its high-latency batch model, instead, is not effective for fast computations or real data analysis.

A widespread tool for Big Data application design is Apache Spark [25], a framework optimized for low-latency tasks. Spark caches data sets in memory and has a very low overhead in launching distributed computations. As stated in Spark website, Spark can "run programs up to 100 times faster than Hadoop MapReduce in memory, or 10 times faster on disk." In multi-step jobs, moreover, Hadoop MapReduce blocks each job from beginning until all the preceding jobs have finished. This can lead to long computation times, even with small data sets. There are other ways to schedule tasks, one of which is *Directed Acyclic Graphs (DAG)*. A graph is used, where the vertices represent the jobs and the edges specify the order of execution of the jobs themselves. Since the graph is acyclic, independent nodes can run in parallel, resulting in a much lower overhead compared to the traditional MapReduce. Spark offers capabilities for building highly interactive, real-time computing systems using DAGs and so is very suitable to implement applications which require an high level of parallelism. Spark is built against Hadoop in order to access *Hadoop Distributed File System (HDFS)*. The key concept beyond Spark is called *Resilient Distributed Dataset (RDD)* [24]. An RDD is a read-only, partitioned collection of records. Data are partitioned across many nodes in the cluster. Fault tolerance techniques are used to avoid data loss due to node failures. Given an RDD we can manipulate the distributed data through operations called *transformations* and *actions*. Transformations consist in the creation of new data set from an existing one, and actions in running a computation on the data set and returning the results to the driver program. We recall that, in the MapReduce paradigm, map is a transformation, reduce is an action. We point out that, in our architecture we will leverage MapReduce for effectively indexing data.

2.2   Evolution of HIS

A HIS [16] is an information system with the aim of capturing, storing, managing or transmitting information related to the health of individuals for contributing to a high-quality and efficient healthcare. Three generations of HIS have evolved in the last decades. The first generation consisted in HIS limited within small facilities, such as departments of hospitals. This type of HIS manages the digitized form of medical documents, such as images or reports created by means of editing programs. A practical example is Radiology Information System (RIS) [17] for storing and managing radiology-related documents. The second generation, born in the 1990s, concerned the integration of such departmental information systems so as to support combined information processing in the hospital. A first example of such integrations was the so-called Electronic Medical Records (EMR), which are legal records created in hospitals and ambulatories, including documents and images, which are consulted by healthcare professionals from a single organization. Another examples is the Picture Archiving and Communication System (PACS) [17], a system for managing and communicating medical images, often integrated with the systems of different departments within the hospital, such as RIS. Such evolution contributed to increase the size of the HIS and the amount and diversity of the exchanged data. In fact, the transition from the first two generations imposes the resolution of technical and syntactical interoperability, i.e., technological and protocol compatibility and diversity in formats of medical data. The DIOGENE project [11], which integrates all patient-related information so as to obtain a seamless communication between hospital actors, is an example of this transition.

The current third generation consists in integrating the hospital-wide HIS so as to form regional HIS, and in federating the these ones so as to have national and trans-national HIS. In this context, Electronic Health Records (EHR) represent a subset of the EMRs issued by each healthcare provider that took care of the given patient during his/her clinical history. These systems permit to share medical information about patients and to have patient-related information following him/her through the various healthcare providers in a given region or country. Practical examples are the Clinical Data Repository/Health Data Repository (CHDR) [12] and the epSOS project. The first one consists of interconnecting all the offices belonging to the Department of Defence and the Veterans Affairs over the overall territory of United States. The second one aimed at designing and developing a service infrastructure supporting the interoperability among every national HIS in several European countries. epSOS is connected to similar initiatives running in the European countries participating to the project, for integrating their regional HIS. The evolution towards HIS of third generation has the consequence of increasing system size in terms of number of interconnected components and amount of exchanged data, but it also exacerbates the interoperability issues to be addressed. Specifically, the transition towards the third generation adds also semantic and business Interoperability, that are common information models/terminology, and common business processes. Figure 1 summarizes the described features.

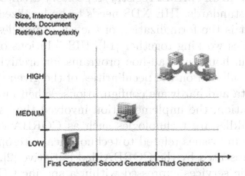

Fig. 1. Systems classification

Interoperability has always been considered as a key challenge to be faced in the described evolution of HIS. since the transition from the first to the second generation, there has been the need of defining standard formats for medical imaging storage and transmission, bringing to the specification of Digital Imaging and Communications in Medicine (DICOM). With the the third generation, as mentioned, there has been the need of specific solutions and guidelines for driving an interoperable HIS interconnection. In the last years, different solutions and guidelines have been proposed and formalized. The first example is represented by the international not-for-profit Foundation called openEHR, which issued a detailed and tested specification for an interoperable HIS platform. Such a vision of openEHR had a significant influence on the development of the emergent healthcare industry standards, such as Health Level 7 (HL7) and CEN EN13606, with recommendations for an interoperable interconnection of HIS. HL7 has defined numerous specifications for enabling interoperability among health applications: among others, two different relevant specifications are Clinical Document Architecture (CDA), based on HL7 v3, which defines the XML schema (format) of exchanged medical documents; the recent Fast Healthcare Interoperability Resourse (FHIR), based on the evolution of HL7 v3, specifies a large, pictorial, representation of medical data in resources. EN13606 represents a subset of openEHR [21], with a specification of the data exchange issues and not for a full federated HIS of the third generation, which is contained in openEHR. HL7 CDA and FHIR support instead syntactic and semantic interoperability by introducing a common model for exchanged medical data. Nevertheless, the history of healthcare interoperability of the last three decades has shown that healthcare standards are not sufficient alone to ensure interoperability. Indeed, they include many if not all the possible situations, thus suffering from various ambiguities and offering many choices that hamper interoperability [13]. To address these issues, the Integrating the Healthcare Enterprise (IHE) initiative has specified some integration profiles, like Cross-Enterprise Document Sharing (XDS). IHE XDS aims at facilitating the sharing of clinical documents within an affinity domain (a group of healthcare facilities that intend to work together) by storing documents in an ebXML registry/repository architecture. In a similar way than the other standards, IHE XDS needs to be localized by specifying an affinity domain, that is the formalization of the set of policies, codes and rules shared by the facilities working together [14]. HIS solutions of the first generation were basically in-house and ad-hoc programs for archiving and retrieving medical documents, tailored on the peculiarities of the given system in terms of type of managed data and hardware configurations. When moving towards HIS of the second generation, the implementation involved the use of well-assessed and -established middleware technologies such as CORBA or DCOM, due to their ability of resolving issues related to technological interoperability. A practical example is represented by the CORBAMed initiative [2], which presents a set of domain-specific services expressed within a specific CORBA domain for the medical environment. Also WS technology is used within the context of the second generation, such as WebCIS described in [22], thanks to the ability of XML-based communication to deal with syntactical interoperability, as proved in [1]. For the third generation, the preferred solution is to use Web Service technology since it has demonstrated a high interoperability capacity and the flexibility to integrate already-existing legacy systems. The previously mentioned standards do not demand a specific technology for implementing federated HIS; however, they recommend SOAP communications for exchanging medical data. This has brought key stakeholders to drive the technological choice towards WS. As a concrete example, we can cite the mentioned epSOS project and CHDR, which are implemented by means of WS. However, the current research has also moved towards different kinds of middleware solutions, such as the Tuple Space-based infrastructure described in [19]. At the moment, the products implemented by these recent research efforts have been scarcely applied in concrete usage. Although XML-based communications resolve syntactic interoperability, they represent only a pre-requisite to the semantic one, and proper additional mechanisms are needed. In the current literature, Semantic Interoperability is typically addressed by means of proper ontologies integrated within the communication system to provide the defining concepts of the given domain [8]. Such a solution has been investigated within the context of healthcare [9], and practically adopted in [19]. Although the foundation ontologies for healthcare have been developed by academic research, none of them have been adopted in concrete applications. In fact, [10] arguments that ontologies for healthcare are not mature solutions, yet. This can be also seen if we study the previously-mentioned standards: only openEHR has specified an ontology to be used in medical data sharing. The most common solutions, i.e., the one adopted by all the other standards and in practical use cases such as epSOS, is to adopt a reference common model for the communications among HIS, as the one specified in HL7 RIM, and a set of mediators for translating from/to such a common model towards the one adopted by each specific HIS.

Securing web services has been an active research topic in the last decade and has been standardized in the OASIS specification called WS-Security [20], which is a composite standard made by combining other different specifications and methods, and specifies two different levels of mechanisms to enforce the provided security level: (1) the first is implemented at the message level by defining a SOAP header that carries out extensions to security; (2) the second is realized at service level to perform higher-level security mechanisms such as access control or authentication. In particular, at the message level we can find two main XML security standard techniques that can be introduced in the mentioned SOAP header extensions: XML Signature and XML Encryption. The former aims at having a small portion of the XML content digitally signed (such element is called digest) so as to provide integrity and non-repudiation for the overall XML content. On the other hand, the latter has the goal to encrypt a part of the overall XML content by using a certain key, which can be public or private according to the chosen encryption strategy. In the case of WS-Security, the SOAP header has a given field, called DigestValue, to contain the digest with indications of the adopted signature method. If encryption is used, the SOAP header has to contain the adopted key, which is itself encrypted by using a proper public key. Besides these two important message-level methods, we have an additional one: Secure

Socket Layer (SSL), which realizes a secure form of the TCP transport protocol, by offering mechanisms for the key agreement, encryption and authentication of the endpoints in a connection-oriented communication. On top of these message-level mechanisms, we can find service-level ones: (i) Security Assertion Markup Language (SAML) is a framework to exchange authentication and authorization information in a request/respond manner when the communication participants do not share the same platform or belong to the same system. The core of this framework is the assertion, expressed with XML constructs, containing the identity of the requestor, and the authorization decisions or credentials. (ii) Extensible Access Control Markup Language (XACML) is used to specify roles and policies used by an access control mechanism to infer the access decisions for users. Different HIS can adopt their own access roles and grants, and XACML is used to exchange such decisions among HIS and to orchestrate their access decisions. (iii) Last, we can find two other specifications: Extensible Rights Markup Language (XrML) and XML Key Management Specification (XKMS). The first is used to express rights and conditions related to the access control (such as expiration times); while the second defines interfaces for the distribution of keys used in XML Signature and XML Encryption.

Security is a key issue in HIS, and the review in [7] provides a complete view of the research efforts spent and achievements obtained so far. As a matter of fact, few works focus on architectures and frameworks, but more focus has been given to qualitative research, modeling and economic studies. Based on these research activities, few prototypes have been realized [15].

3 Advanced Data Search

3.1 Complex Search

Nowadays, the availability of huge amounts of information calls for proper solutions to the complex search problem. To this end, search engines have been proposed since the early stage of Internet. However, results returned by search engines are often quite far from the expected query answers from a user viewpoint. Indeed, search results can be improved by building a custom map that, based on the initial query results, tries to learn additional knowledge about data being queried by iterative refinement of search dimensions and parameters. Figure 2 shows the above mentioned scenario.

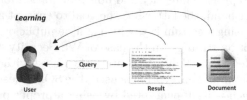

Fig. 2. Learning by results

In this scenario, the type of query being performed plays a crucial role. Unfortunately, this process is suitable only for simple search of well-defined terms. On the contrary, dynamic learning by exploratory research cannot be performed by this naive process. Obviously enough, for well defined queries, a search engine like Google, is able to provide correct results in a few milliseconds[1].

However, in some cases users do not know exactly how to find the desired information about an object or a service (e.g. a book or a restaurant). In this case, the model depicted in Fig. 3 is more suitable.

Fig. 3. Amazon search

More in detail, Amazon-like search tools, feature product categorization and recommender systems, thus making the user search experience quite interactive and iterative. In a sense, intermediate results guide users to a better definition of target information. Furthermore, search engines usually allow non-structured queries (referred as "ranked retrieval") whose results are sorted according to some relevance criteria w.r.t. the target search. As a matter of fact, these queries are easier to pose by users compared to boolean expressions, but they can produce low quality results.

In order to overcome this limitation, some categorization service like Yahoo!Directory, exploits context information[2]. More in detail, directory contents are hierarchically organized in order to guide users through a subset of documents potentially related to information being queried, thus limiting the possibility to input free text queries. In this respect, users re-think and refine their needs by learning the adjustments to the search being performed by exploiting the available choices. To better understand how directory navigation works, we resurge to accommodation booking portals analogy. Indeed, those portals offer a hierarchical navigation systems, i.e. from the home page, user can choose

---

[1] As a matter of fact, due to its quick result presentation, many users go through Google even if they exactly know the URLs of the resources they are interested in.

[2] Yahoo!Directory is no longer active since 2014, however it is worth mentioning as it was one of the first services for massive assisted browsing.

the desired country, then s/he can specify the city and finally the type of structure s/he is interested in. This navigation model suffers a great limitation due to
taxonomy specification. Indeed, taxonomy specified by the service designer may
not meet user needs. A solution to overcome the above mentioned limitations
is the implementation of *faceted* navigation that helps users in the information
"surfing" process.

4   A Big Data Architecture for Supporting EHR

In this section, we describe the overall architecture of our proposal for assisting
medical data search. Our goal is to provide users a flexible tool for assisted text
search, that is interactive, scalable and dynamic in order to easily connect and
integrate all the nodes of the healthcare infrastructure. To this end, we exploited
several indexing and data management strategies that allow us to cope with
high volume, heterogenous and burst information. Figure 4 shows our system
architecture.

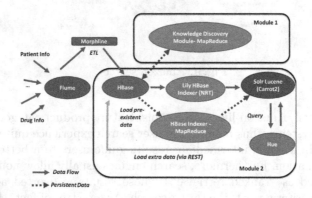

Fig. 4. System architecture for big data search (Color figure online)

In order to guarantee maximum implementation flexibility, we exploited several open source tools as Apache Hadoop[3], Flume[4], HBase[5], Solr[6], Lily HBase
Indexer[7] and Hue[8].

Data are collected in our system from heterogeneous sources and they arrive
in a streaming way. In order to properly manage these data, we implemented
some specialized *Crawling* services that are closely tied to the data set being

---

[3] http://hadoop.apache.org.
[4] https://flume.apache.org/.
[5] https://hbase.apache.org/.
[6] http://lucene.apache.org/solr/.
[7] http://ngdata.github.io/hbase-indexer/.
[8] http://gethue.com/.

collected. As data are crawled, we collect them by *Flume* module (note that the blue arrows in Fig. 4 refer to data flows arriving at different rates from multiple sources). This module will host our staging area, as it is a reliable and distributed service designed to efficiently collect, aggregate and forward huge amounts of data for later storage in a permanent repository. Furthermore, it is well suited for dealing with data streams, as it provides fault tolerance by an easily configurable reliability mechanism that are mandatory for managing healthcare data. The latter feature has been profitably exploited for dealing with data inconsistency due to null or missing values that could arrive from multiple data sources.

Once data are gathered by Flume module, they undergo through an "on the fly" ETL (Extraction, Transformation and Loading) process performed by *Morphline* module. This module is devoted to data cleaning and data mapping on the column set in the datastore. The output of this step is a cleansed data flow on top of which our analysis takes place. Collected information are sent to our big data storage layer, implemented by *HBase*, that is devoted to data storage.

As stated above, our goal is to improve full-text search. To this purpose, we exploit *Apache Solr* tool for our discovery task (more precisely we used the *SolrCloud* implementation). The rationale for exploiting *Solr* is twofold: 1) it allows effective and efficient searching for keywords appearing in any column that has been previously indexed and 2) it allows to faster display documents ranked by their relevance w.r.t. the query being issued. Moreover, *Solr* provides several useful presentation features as: field facets, range queries and pivot facets that allow a proper organization of the results to be shown to the user. Those operators can be also fruitfully exploited for providing users the classical on line analytical processing operators as slice & dice, drill-down, roll-up and pivoting. In this respect, *Solr* has been proven to be an excellent real-time analysis engine for text documents (like user queries and suggestions exploited in our system). Consider, as example, web site logs: *Solr* can easily indexed them in order to execute (time-stamped) range queries for a given (set of) keyword(s). Moreover, it is possible to build the information graph containing aggregate information, such as the growth over time of registered users or transactions grouped by type. We exploit these information in our system for providing better suggestions. As users interact with our system, e.g. by searching new information or by posting new documents related to a disease, new data are collected by the storage layer. Based on data arrival rate, we schedule offline clustering of the whole dataset (e.g. after a burst of tuples is collected) in order to better organize data and for boosting the indexing strategy assisted by ad-hoc MapReduce functions.

In order to properly display search results, we exploit *Hue* features. The latter, is a software offering a customizable user friendly interface.

5   Case Study: A Nationwide System

In this section we present our case study, i.e. a nationwide system. In particular, we will first show the law requirements to be met.

5.1 Law Requirements

A recent Italian decree establishes security and organizational requirements that regional EHR systems (EHR-S) must provide:

- Implementation of a set of functions based on a shared functional model. Specifically, a common functional model for EHR-S, obtained localizing the HL7/ISO EHR-S FM standard, has been defined by an interregional initiative, comprising regional representatives, some government agencies and associations (i.e. HL7 Italy), and research institutions. The functional model defined specifies, in a structured and integrated way, a set of business functions for the EHR, delegating implementation details about the realization of interoperable EHR-Ss. The profile, published by HL7 Italy, has been defined through an analysis of the existing laws, rules, work processes, and actors involved in the use of an EHR-S.
- Development of an enabling platform able to connect all the healthcare facilities distributed on the regional territory, in order to enable users to search, insert, and retrieve documents within their purview.
- Realization of services aiming at collecting health documents generated from health professionals. The mandatory documents that each EHR-S has to be able to handle are Patient Summary (PS) and Laboratory Report. To this scope, the software applications used by the health professionals have to be integrated with the regional platform.
- Implementation of a service that enables a health professional to identify a patient before he/she requests the system to access clinical documents.
- Integration of consent management services with the regional platform, in order to satisfy the legal constraint according to which health documents can be uploaded and consulted only if the patient has provided, respectively, two kind of informed consents: one aimed at making the patient able to express his/her intention to allow health professionals registering documents into his/her EHR; another allowing the patient enabling the consultation of his/her documents to all the health professionals that have the roles he/she authorized.
- Implementation of access policy management integrated with all the business services. EHR-Ss must satisfy the will of the patient, which is expressed through policy policies. It is therefore necessary to establish strict authentication and authorization policies for documents access.
- Implementation of interoperability services integrated with the regional platform in order to make this one able to interact with other regional platforms for (i) searching, (ii) retrieving, and (iii) registering health documents.

5.2 Experimental Infrastructure

An experimental interoperability infrastructure conformed to the Italian norms and technical specifications has been implemented with the scope of using the proposed approach for accessing EHR document managed by some Italian Regions (referred in the following as Region 1, 2 and 3).

The experimentation consists in enabling regional HIS to exchange medical documents related to some patients, available at the various healthcare facilities. In particular, the HIS of Region 1 and Region 2 have a similar architectural model that we cannot show in detail for regulatory issues. The operations experimented enable physicians to query and retrieve health documents of a patient which are available in another region, e.g. because in the past he/she has benefited from a health service in this region. The interconnection of the interoperability infrastructure with the regional HIS of the three regions has required a set of actions, described below.

The platforms of the Region 1 and Region 2, which share the same architectural model, have been integrated at the same way with the big data infrastructure. First, an Access Interface and an Indexing Strategy components have been deployed at each regional node of the infrastructure; in particular, the Indexing Strategy component interacts with the storage layer. Second, several instances of the Document Manager components have been deployed at a set of healthcare facilities. Such components have been integrated with the information systems of such facilities by means of wrappers able to translate the standard protocols with the ones used by the legacy systems. The actions performed for the integration of the HIS of the Region 3 with the big data infrastructure are: (1) development of a wrapper able to interconnect the Index Strategy component with the registry of the regional HIS, with the aim to translate the language used to represent the metadata of the shared information model with the one used by the local HIS; (2) implementation of a wrapper capable of interacting the Access Interface component with the legacy repositories of the healthcare facilities.

The healthcare metadata and documents related to a patient available in the regions different from those where a patient resides can be performed in two steps. The first step consists of a simple search: a user (e.g., a general practitioner) sends a query to the regional HIS, which propagates it to the Indexing module; it (i) makes the query to its own data store, (ii) interacts with the overall Indexing module (that is automatically maintained by our big data infrastructure) of the other regions, which executes the query to their registries, (iii) aggregates all the metadata results, and (iv) returns the results to the user. The second step is a document retrieval: the user selects a document s/he wants to obtain and sends a request to the regional HIS, which forwards it to the Access Interface component of the region containing the document; this one retrieves the document by communicating with the HIS of the region where it is deployed. It is worth noting that the regional platforms receiving the query and retrieve requests from the other regions have to be adequately processed and verified. The verification process consists in two phases: access control and information availability.

The first phase (access control) has the aim of verifying that the requesting user has the right to access the health documents he/she demanded. The access control system, based on an XACML architetcure, performs the following steps:

1. a Policy Enforcement Point (PEP) intercepts the message sent from the inter-operability service of another region;
2. the PEP analyzes the claims transmitted in the messages in order to verify that they have all the attributes necessary (like the role of the user, the purpose of use, etc.). In particular, the claims are represented as assertions according to the SAML standard;
3. the PEP verifies the validity of the digital signatures of the SAML assertions;
4. the PEP controls if the patient has provided the opportune consents;
5. the PEP forwards the request to the Policy Decision Point (PDP);
6. the PDP provides the final decision (that is, permits or denies the access to the service). The decision is taken on the basis of the privacy policies established by the patient: the most important aspect to verify is that the role of the user contained in the assertions has the right to access the service. The final decision is transmitted to the PEP;
7. if the decision is a PERMIT, the PEP forwards the message to the appropriate service on the basis of the operation requested (that is, query or retrieve); if the decision is a DENY, the PEP returns an error message to the user.

The second phase is realized in case of a PERMIT response from PDP and is realized by the service invoked. In case of query, the service interacts with the registry in order to verify if there are entries that meet the search criteria indicated by the user: all the metadata satifying such criteria are returned to the user. In case of retrieve, the service interacts with the appropriate repository to obtain the health document satisfying the request: the document, if available in the repository, or an error message is sent to the user.

The quality of the search and retrieval operations have been tested with an experiment scenario consisting in a user in a given region that wants to obtain a document of a patient from a different region. Another experimentation has been made considering an intra-regional scenario, i.e., the user in a given region intends to obtain a document of a patient from the same region. With regards to the first scenario, we have performed from one region about 125 requests of search and retrieval operations of document identifiers randomly chosen among those hosted by the other two regions. The system returns two versions for each medical document requested: one according to the XML-based HL7 CDA Rel. 2.0 format of about 50 KB and the other one in PDF/A format of about 270 KB. We had the 100 % success rate (in terms of appropriate results provided to the user) for searched documents. With regards to the second scenario, the results have shown that almost half of the time is needed to retrieve the searched documents as we avoid the communication cost among regional nodes.

6 Conclusions and Future Works

This work presents a big data based architectural model aiming at enabling access to regional EHR systems, which have to be developed or revised according to recently issued specific Italian laws. The work represents an important first step in the process of digitizing the national EHR system. The proposed model is turning out to be successful for both Regions that have already started an e-health process and Regions that are still in a start-up phase. The efforts which have been made so far help the organizations to overcome the main difficulties to treat large amount of health data, highlighting the benefits that automated processes could bring in terms of time efficiency and care effectiveness. The solutions described in this paper are quite flexible as, on the one hand, they provide a standardized approach to ensure interoperable access to health data for processing. Anyway, due to regulatory issues we are still at the early stage of our experimental assessment as we need to overcome some of the above mentioned legal problems in order to fully exploit the potential of the proposed system. Thus, we are planning to define specific architectural components with the aim of performing anonymization operations when necessary.

References

1. Synapses/SynEx goes XML. Studies in Health Technology and Informatics, IOS press (1999)
2. (2001). http://healthcare.omg.org/Roadmap/corbamed_roadmap.htm
3. Big data. Nature., September 2008
4. Data, data everywhere. The Economist., February 2010
5. Drowning in numbers - digital data will flood the planet - and help us understand it better. The Economist., November 2011
6. Agrawal et al., D.: Challenges and opportunities with big data. A community white paper developed by leading researchers across the United States., March 2012
7. Appari, A., Johnson, M.E.: Information security and privacy in healthcare: current state of research. Int. J. Internet Enterp. Manage. **6**(4), 279 (2010)
8. Bittner, T., Donnelly, M., Winter, S.: Ontology and semantic interoperability. Large-Scale 3D Data Integration. CRC Press, London (2005)
9. Blobel, B., Oemig, F.: What is needed to finally achieve semantic interoperability? IFMBE Proc. **25**(12), 411–414 (2009)
10. Blobel, B., Kalra, D., Koehn, M., Lunn, K., Pharow, P., Ruotsalainen, P., Schulz, S., Smith, B.: The role of ontologies for sustainable, semantically interoperable and trustworthy ehr solutions. In: Medical Informatics in a United and Healthy Europe - Proceedings of MIE 2009, The XXIInd International Congress of the European Federation for Medical Informatics, Sarajevo, Bosnia and Herzegovina, Agust 30 - September 2, 2009, pp. 953–957 (2009). http://dx.doi.org/10.3233/978-1-60750-044-5-953
11. Borst, F., Appel, R., Baud, R., Ligier, Y., Scherrer, J.: Happy birthday diogene: a hospital information system born 20 years ago. Int. J. Med. Inf. **54**(3), 157–167 (1999)
12. Bouhaddou, O., Warnekar, P., Parrish, F., Do, N., Mandel, J., Kilbourne, J., Lincoln, M.J.: Exchange of computable patient data between the department of veterans affairs (va) and the department of defense (dod): terminology mediation strategy. J. Am. Med. Inf. Assoc. **15**(2), 174–183 (2008)
13. Dogac, A., Laleci, G.B., Aden, T., Eichelberg, M.: Enhancing ihe xds for federated clinical affinity domain support. IEEE Trans. Inf. Technol. Biomed. **11**(2), 213–221 (2007). http://dx.doi.org/10.1109/TITB.2006.874928

14. Dogac, A., Laleci, G.B., Kabak, Y., Unal, S., Heard, S., Beale, T., Elkin, P.L., Najmi, F., Mattocks, C., Webber, D., Kernberg, M.: Exploiting ebxml registry semantic constructs for handling archetype metadata in healthcare informatics. Int. J. Metadata Seman. Ontol. **1**(1), 21–36 (2006). http://dx.doi.org/10.1504/IJMSO.2006.008767

15. Esposito, C., Ciampi, M., De Pietro, G., Donzelli, P.: Notifying medical data in health information systems. In: Proceedings of the 6th ACM International Conference on Distributed Event-Based Systems. pp. 373–374. DEBS 2012, NY, USA. ACM, New York (2012). http://doi.acm.org/10.1145/2335484.2335528

16. Haux, R.: Health information systems past, present, future. Int. J. Med. Inf. **75**(3–4), 268–281 (2006)

17. Huang, H.K.: PACS and imaging informatics. Wiley-Liss, Hoboken (2004)

18. Masciari, E., Mazzeo, G.M., Zaniolo, C.: Analysing microarray expression data through effective clustering. Inf. Sci. **262**, 32–45 (2014). http://dx.doi.org/10.1016/j.ins.2013.12.003

19. Nixon, L.J.B., Cerizza, D., Valle, E.D., Simperl, E., Krummenacher, R.: Enabling collaborative ehealth through triplespace computing. In: Proceedings of the 16th IEEE International Workshops on Enabling Technologies: Infrastructure for Collaborative Enterprises, pp. 80–85. WETICE 2007, IEEE Computer Society, Washington, DC (2007). http://dx.doi.org/10.1109/WETICE.2007.140

20. Nordbotten, N.A.: Xml and web services security standards. IEEE Commun. Surv. Tutorials **11**(3), 4–21 (2009). http://dx.doi.org/10.1109/SURV.2009.090302

21. Schloeffel, P., Beale, T., Hayworth, G., Heard, S., Leslie, H.: The relationship between cen 13606, hl7, and openehr (2006)

22. Sittig, D., Kuperman, G., Teich, J.: Www-based interfaces to clinical information systems: the state of the art. In: Proceedings of the AMIA Annual Fall Symposium, pp. 694–698. CRC Press, London (1996)

23. White, T.: Hadoop: The Definitive Guide, 1st edn. O'Reilly Media Inc, Sebastopol (2009)

24. Zaharia, M., Chowdhury, M., Das, T., Dave, A., Ma, J., McCauley, M., Franklin, M.J., Shenker, S., Stoica, I.: Resilient distributed datasets: A fault-tolerant abstraction for in-memory cluster computing. In: Proceedings of the 9th USENIX conference on Networked Systems Design and Implementation, pp. 2–2. USENIX Association (2012)

25. Zaharia, M., Chowdhury, M., Franklin, M.J., Shenker, S., Stoica, I.: Spark: Cluster computing with working sets. In: Proceedings of the 2Nd USENIX Conference on Hot Topics in Cloud Computing, pp. 10–10. HotCloud'10, USENIX Association, Berkeley, CA, USA (2010)

Integrating Open Data on Cancer in Support to Tumor Growth Analysis

Fleur Jeanquartier[1]([✉]), Claire Jean-Quartier[1], Tobias Schreck[3],
David Cemernek[1], and Andreas Holzinger[1,2]

[1] Holzinger Group, Institute for Medical Informatics, Statistics and Documentation,
Medical University Graz, Graz, Austria
{f.jeanquartier,c.jeanquartier,d.cemernek,a.holzinger}@hci-kdd.org
[2] Institute of Information Systems and Computer Media,
Graz University of Technology, Graz, Austria
[3] Institute of Computer Graphics and Knowledge Visualisation Graz,
University of Technology, Graz, Austria
tobias.schreck@cgv.tugraz.at

Abstract. The general disease group of malignant neoplasms depicts one of the leading and increasing causes for death. The underlying complexity of cancer demands for abstractions to disclose an exclusive subset of information related to the disease. Our idea is to create a user interface for linking a simulation on cancer modeling to relevant additional publicly and freely available data. We are not only providing a categorized list of open datasets and queryable databases for the different types of cancer and related information, we also identify a certain subset of temporal and spatial data related to tumor growth. Furthermore, we describe the integration possibilities into a simulation tool on tumor growth that incorporates the tumor's kinetics.

Keywords: Open data · Data integration · Cancer · Tumor growth · Data · Visualization · Simulation

1 Introduction

Interactive data integration, data fusion and, first and foremost, the selection of datasets is a key research direction to enable knowledge discovery in health informatics generally, and bioinformatics and computational biology specifically [1].

Our aim is to link publicly and freely available data on cancer to an enhanced version of our recently presented tool on tumor growth [2]. Thereby, we list open databases providing datasets on the different types of cancer and collect related information. The datasets are examined for growth-related parameters and subsequently integrated into a simulation tool on modeling neoplasms. This simulation on neoplasia comprises abnormal tissue growth such as benign and malignant tumors. Additional text-based information and non-growth-relevant data is scanned and revised for accessory visualization features.

© Springer International Publishing Switzerland 2016
M.E. Renda et al. (Eds.): ITBAM 2016, LNCS 9832, pp. 49–66, 2016.
DOI: 10.1007/978-3-319-43949-5_4

We further describe and sketch possibilities for integration and visualization of cancer-related data into our recently presented simulation and visualization tool on tumor growth [2]. The Web tool is based on the implementation of the Cellular Potts Model (CPM) and Cytoscape, that is available at https:// github.com/davcem/cpm-cytoscape. We present an integrative approach to cancer research. The study rests upon the idea of enhancing the tumor growth simulation by integrating multiple genuine data.

First, we introduce the topic of open data for research in general and on cancer in detail. Further, we recap the biological settings for cancer modeling. We approximate and appoint open datasets on cancer involving tumor growth information by considering temporal and spatial aspects. And, we discuss their feasible incorporation into an online simulation. We proceed with a summary on the key challenges for embedding open data to our cancer simulation. We thereby suggest that an integrative approach is key to understanding cancer.

2   Related Work

2.1   Open Data for Scientific Research

There is a strong trend towards an increasing number of freely available datasets becoming available in many domains, including scientific research. The idea of open data is to provide unrestricted access for sharing, validating, reusing and merging relevant data to advance scientific research. Several works already show that new opportunities arrive with the increasing amount of open data. The so-called *Fourth Paradigm* [3] envisions data-driven research by widened access to open data for common good.

While open data provides opportunities, there are challenges associated with the provision, discovery and usage of open data. Typically, relevant content needs to be retrieved by researchers. Then, data from different sources of possibly heterogeneous data regarding data type, quality, and resolution need to be integrated for joint analysis.

Interactive visualization can help to explore and related data during the discovery process. Domain- as well as application-specifics need to be taken into account to choose the right visualization tool for supporting search and exploration in general data exploration [4–6]. In previous work, approaches for discovery of relevant data in research data repositories based on exploration and visual querying have been proposed. The VisInfo system [7] allows to query for content in large time series databases. Often, content needs to be related to metadata. In [8] data patterns are correlated with metadata, for enhanced exploration. Visual search for bivariate data has been addressed in [9] using features obtained from scatter plot representations of input data. In absence of example queries from real data, user sketching of patterns can be useful, if appropriate similarity functions can be obtained [10]. Besides exploration, visual-interactive approaches can also be useful for the effective semi-interactive integration of heterogeneous data sources, which is a primary requirement in many open data analysis projects [11].

More specifically regarding the medical domain, we recently compared methods for visualizing and analyzing data in online proteomics databases. Only a few available tools meet the needs for interactive visual analysis [12].

Increasing data availability is not only considered as an opportunity but also new issues arise. Challenges of data integration in the biomedical sciences include determining available and usable data, completeness, re-use for novel approaches for data discovery and exploitation [1,13].

2.2 Open Data in Cancer Research

Biomedical data comes in many guises [1]. Initiatives are already fostering open-access research for improving patient care. There are several freely accessible web portals, yet, providing exploration support for cancer genomics due to increasing efforts in the area of Bioinformatics regarding genomic data handling [14–22]. For example, challenges in normalizing clinical drug data have been illustrated while using open access druggable genome datasets for target discovery in the context of cancer therapeutics [23].

With regard to imaging data there are several online resources providing several million cancer images, which are partly public, partly protected. Available imaging data includes computed tomography, magnetic resonance and other images. De-identification scripts support moving more and more images on public servers [24].

Text mining for literature curation is common for omics data [25]. Summaries of fundamental concepts for text mining in cancer research are mainly concerned on relation extraction mechanisms such as identifying protein-protein, gene-gene or gene-disease relations [26]. Text mining has already been combined with manually curated data for data integration in the context of disease-gene associations [27]. Several open access literature resources exist to apply text mining for finding suitable disease data. However, text mining in biomedical literature is more sophisticated than for clinical data [28]. Only a few databases provide information on cancer incidences and statistics. Movements come from the American Cancer Society and the World Health Organization [29–31]. Data protection regulations and privacy is one of the obstacles to tackle to providing open data for biomedical research [33,34] There are approaches for space-time analysis and visualization related to cancer, but they deal with population data such as location and age [35].

Sophisticated integrative analysis tools for cancer are yet to be found [36]. Online available disease ontologies help understanding the relationships of cancer terms and foster communication and exchange [37,38].

To our knowledge, there is no approach to identifying tumor growth related open data. We therefore focus on identifying temporal and spatial entities within available cancer data.

2.3   Biological Background

There are two basic biological phenomenons which play essential roles in the disease of cancer. First, spontaneous mutations occur naturally and frequently within all cells [39]. Secondly, normal cells can undergo programmed cell death, so-called apopotosis, with time. In some cases however, such mutations can have an effect on cellular functions. Tumor cells are characterized by a change in the proliferative capacity. Malignancy can be developed if mutations lead to the inhibition of apoptosis or excessive proliferation and could further end in differentiation. Tumors can look and function similar to normal cells. Benign masses of tumor cells are normally localized. They only become problematic if space is limited or keep producing hormones in excess [40]. Malignant tumor cells become more serious. They do not only grow more rapidly but they can also invade other tissues and parts throughout the body. Parameters that relate to the specified aspects in tumor growth are of particular importance for modeling cancer. Since mutations are the onset of cancer, open data is concentrated on genetic data. Still, in order to combat the disease relational information has to be retained.

3   Approach

Our approach is to study open datasets for querying and relating interaction data to (gene classified) cancer diseases. The goal is to extend an existing framework for simulating and visualizing tumor growth [2] by integrating a selected subset of spatial and temporal data for supporting exploration and sense-making. To achieve this goal several data integration steps are necessary. Most important, available data has to be identified and examined for relevance.

3.1   Relevance to Tumor Growth?

We focus on summarizing and picking specific information on tumor growth. Presently, there are no web-resources providing exclusive data on tumor growth. So, relevant information has to be isolated from an abundance of data in matters of cancer research. We aim to gather cancer-relevant data in regard to spatial and temporal criteria in particular.

Temporal and spatial characteristics on tumor growth can be influenced by several factors, such as gene regulation or mutations as well as drugs and other inhibitors or promoters. In cancer, the balance between growth promoting and inhibiting factors is shifted towards proliferation. The underlying signal-transduction pathways are complex biological processes involving several key steps as well as mediators which are dynamically and differentially regulated. The influencing factors have to be recognized and parameterized in order to be integrated into the simulation.

We are equally interested in statistical assessment of growth kinetics from various tumors and cancer subtypes, as well as incidence reports on isolated case reports. Notably, entity relationship descriptions and interaction data in regard to tumor growth characteristics are of relevance and primary focus.

Previous studies on tumor growth prediction could be likewise included. In order to enhance the cancer modeling tool, we aim to provide a comprehensive simulation comprising growth characteristics of various kinds of tumors. Most studies on predictive cancer modeling focus on the kinetics of various cancer diseases. We try to collect and capture the specifics of several tumor types and to likewise broaden and refine the visualization approach tumor growth analysis.

4   Results

We present an overview of available cancer-related open data. We categorize identified datasets corresponding to the content types that can be found with respect to cancer research. The study shows that genomic data as well as imaging data is increasingly available. But, explicit information on temporal and spatial aspects are hardly found. Text mining in incidence reports and open access publications have to be taken into account in order to find suitable data for tumor growth simulation. Furthermore, we describe the integration of a subset of open data related to tumor kinetics, temporal and spatial data in particular, into an existing tumor growth simulation user interface that is freely online available via github.

4.1   Overview of Available Data

We categorize online available information from cancer research under 5 different categories. First, many datasets provide **genomic data**. Secondly, **incidence data** can be analyzed and downloaded from several portals. Third, there are large archives consisting of **imaging data**. Fourth, there are several databases that consist of **disease associations** such as disease ontologies. Last but not least, open access databases provide a comprehensive list of **literature data** for text mining.

By considering content quality, license information and access possibilities for each of the listed entries, we chose a subset that satisfied the needs for free non-commercial usage as well as data relevance. Table 1 lists facts about the identified databases regarding its data category relation.

Starting with a review of currently available cancer genomic databases for research [41], our search strategy included systematically examining lists of databases of cancer-related data presented at metasites found via online search. Therefore, we iteratively extended a table of cancer related databases until we arrived at a comprehensive list of databases that we are summarizing below. We examined available databases and included information about access possibilities as well as descriptions about the provided data type/category, the data's coverage, whether download of data as well as a web API is provided, license information and last but not least studied optional input and output entities.

Table 1. Statistics about list of non-filtered databases

Category	# Identified databases	# Chosen databases	Possibilities for spatial data	Possibilities for temporal data
Genomic data	15+	9	–	–
Imaging data	6+	5	✓	–
Incidence data	6	4	–	✓
Disease associations	6	3	✓	✓
Literature data	2+	2	✓	✓

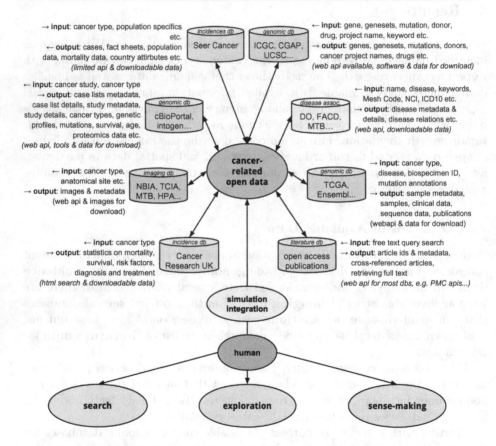

Fig. 1. Overview of cancer databases for integration

Therefore, next to the availability of spatial and temporal data, we further differentiate between possible input and output. Figure 1 shows an overview of our approach. The input and output is being summarized. The node's color corresponds to the data's category.

Table 2. Summary of examined databases that may be suitable for the task of data integration

Category/Name	Abbreviation	Data access	Ref.
Genomic data			
The Cancer Genome Atlas - Data Portal	TCGA	REST, download	[16]
cBio Cancer Genomics Portal	cBioPortal	REST, download	[15]
NCI's Cancer Genome Anatomy Project	CGAP	download	[43]
International Cancer Genome Consortium - Data Portal	ICGC	REST, download	[19]
United States Cancer Statistics - Cancer Genomics Browser	UCSC	download	[16]
Catalogue of somatic mutations in cancer	COSMIC	REST, download	[18]
Integrative Onco Genomics	INTOGEN	download	[20]
Integrative Genomics Viewer	IGV	download	[21]
Many more general genome databases such as Ensembl	ENSEMBL	REST, download	
Imaging data			
The Cancer Imaging Archive	TCIA	REST, download	[24]
CancerData.org - Sharing data for cancer research	CancerData	download	[45]
Mouse Tumor Biology - Database	MTB	download	[44]
National Biomedical Imaging Archive	NBIA	REST, download	[24]
Many more such as the Human Protein Atlas	HPA	download	
Incidence data			
WHO Cancer Mortality Database	WHOdb	download	[46]
Center for Disease Control and prevention - Cancer Data and Statistics	CDC	download	
Surveillance, Epidemiology, and End Results - Program	SEER	download	[30]
Cancer Incidence in Five Continents	CI5	download	[31]
Disease associations			
Diseases Ontology	DO	REST, download	[37]
Mouse Tumor Biology - Database	MTB	download	[44]
NCI Thesaurus	NCIt	REST, download	[38]
Literature data			
PubMed Central	PMC	REST, download	[26]
Europe PubMed Central	Europe PMC	REST, download	[32]

Table 2 lists all examined databases providing cancer-related content as download that is free for non-commercial, scientific purposes, sorted by category.

The summarizing table shows only a small subset of examined resources due to the fact that several licensing issues as well as quality issues such as deprecated data that has not been maintained for years have been identified during our research. We also observed that several data portals make use of others, e.g. the Disease Ontology's cancer project includes several mappings from other databases, especially genomic data. The "+" in the column of identified databases within Table 1 implies that more databases could be found but are already included within other databases. To that effect, the databases' peculiarities also include data coverage such as databases that cover other databases' contents as well. Due to that reason, we chose to use only the largest two archives of biomedical literature data for further literature mining.

4.2   Literature Mining

We conducted a search for some tumor growth related terms to test the suitability of literature databases for finding data to be integrated. PubMed has been reported to be one of the best biomedical publication archives [26]. Therefore, we chose to conduct some mining within the two public archives of biomedical and life sciences literature, "Europe PMC" and "Pubmed Central" (PMC). Additionally, we made use of an information retrieval tool for biological literature called "Textpresso" [42]. Example queries are summarized below.

Table 3. Example queries for text mining

Database or tool	Query for "abnormal cell growth"	Query for "tumor growth"	Query for "tumor cell growth"	Query for "neoplasm"
Textpresso	111 matches, 33 documents	3891 matches, 926 documents	37072 matches, 6519 documents	3990 matches, 2000 documents
Europe PMC	1399 matches, 277 open access	121435 matches, 35174 open access	12555 matches, 4089 open access	4076094 matches, 436216 open access
PMC	1389 matches	98822 matches	13557 matches	2837065 matches

Making use of specific text mining tools is favored over literature mining for finding most relevant results and presenting sets of results. E.g. highlighting matching sentences is crucial to a fast scan through results and the identification of relevant information.

4.3   Data Processing

Most online portals provide free access to the data available as downloadable content, some accompany web interfaces such as web services for direct access too. In each case further data processing steps are necessary to respond to the needs of (visual) data mining and integration into the existing user interface.

Most genomic data portals already provide entity relationship (ER) diagrams for documentation of available data entities and relations. However, we focused on finding temporal as well as spatial tumor growth data and were not able to identify explicit information about those aspects within available cancer genomic data. Further mining techniques have to be taken into account to accomplish the task of finding suitable information about specific growth impact on cancer disease-gene associations.

As a starting point for data integration we created a set of different growth functions by literature curation. We collected data points for comparing discrete growth functions for tumor growth, vascularization inhibition and cell density inhibition on growth. Data points come from three different publications found via PMC and is summarized in Fig. 2 [47–49].

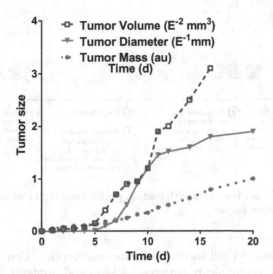

Fig. 2. Literature-curated discrete tumor growth - data samples: various tumor types, determined growth in tumor size, given in miscellaneous units, over time, presented in days.

4.4   User Interface Extensions

Cpm-cytoscape is a tool for scientific simulation and visual analysis of tumor growth. The web application makes use of the CPM for modeling tumor growth. The CPM is a popular lattice-based, multi-particle cell-based model that has been used for modeling tumor growth in a wide area. The tool incorporates a novel graph-based visualization approach [2]. Figure 3 shows an annotated screenshot of the existing user interface, describing the different interaction and visualization possibilities of the tool's user interface.

The tool's framework integrates visualization features for analysis via JavaScript and HTML. A Converter Class allows for extending the data objects

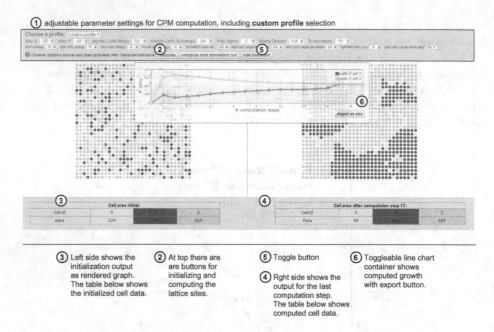

Fig. 3. Overview of User Interface with custom profile showing kinetics and cell sorting after several simulation steps

that represent simulated cell sorting and kinetics. Another Converter allows for processing data to communicate between backend and frontend. This Java Class maps the graph data from the modeling computation to the format needed by the visualization renderer in the frontend. Such converter classes are easily extendable and support integrating additional information. The simulation and its several computation steps are started via Representational State Transfer (REST) calls, while the user interface displays response information both within the graph visualization as well as in an overlay as simple Line diagram. Details on its usage and implementation can be found on the project's github page [2].

Profile Specific Simulation and Visualization. The first implemented extension to the user interface is the ability to provide "profiles" for running simulations under different configurations. The simulation can be started with the help of choosing a profile or specifying a custom profile. Figure 3 shows a completed simulation for a custom profile. The profile extension is a good example of extending the user interface neatly and encapsulated. A separate JavaScript function call via changeProfile() is located in an separate extension. Each profile for selection is represented as JSON file for easy maintenance. The profile can be selected via a dropdown (Fig. 4). The parameter settings that are available via JSON files can be replaced with a dynamic function that communicates with another server to get all the various parameter settings.

Fig. 4. Screenshot of profile selection possibilities

Until now we did not find any database that holds all the data needed to have different complete configurations to run a simulation, therefore we are providing static configuration files to try out different settings that have found via manual literature mining. However, this extension is a good example to start the task of data integration and can be further extended as soon as a suitable dataset is available.

Presenting Details on Cell Nodes. The visualization of cell sorting and kinetics is based on a graph. Each node is representing a so called "cellular brick" of a cell. A cell is a set of 0 to n cellular bricks with the same cell-index, while each cell *sigma* is of a specific cell-type τ. Until now, we only differentiate between proliferating tumor cells and healthy cells as distinct cell types, with different growth rates and volume constraints for each type, rendered as colored nodes. Thirdly, we use grey nodes to represent the extracellular matrix (ECM). Additional information on nodes can be provided via context menu. According to the node's cell-index $\sigma_{i,j}$ additional information about the associated cell-type can be shown, while proliferating tumor cells are called "dark" cells and the other healthy cells are called "light" cells. Cells with $\sigma_{i,j} = 0$ represent the ECM, visualized as grey nodes. Cells with odd $\sigma_{i,j}$ represent the "dark" cells and are visualized as dark red colored nodes. The other cells with an even $\sigma_{i,j}$ show "light" cells and can be recognized by the lighter blue to green colored nodes.

Search for Reports on Related Diagnosis and Treatment. Text-based search within an existing incidence data provides exploration of similar cases, diagnosis, treatment as well as other possible relations. Figure 5 shows a mock-Up of a simple integration. As starting point we just link to additional information. However, a tight integrative approach would be adding further data to the computation of the several simulation's steps. Taking additional information into account such as drug information that has impact on growth could then be presented as uncertainty visualization as sketched in Fig. 6.

Direct Inclusion of Time-oriented Data for Growth Simulation. An ultimate goal is to include information not only on existing related incidences but far more information on drugs and other inhibitors or promoters to be integrated directly into the computation process. In particular time-oriented data as we see in the simple line diagram showing the growth of different celltypes supports integration of additional information to be visualized for further exploration and

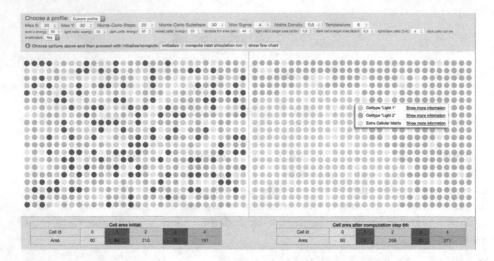

Fig. 5. Screenshot, showing additional information for cell nodes (Color figure online)

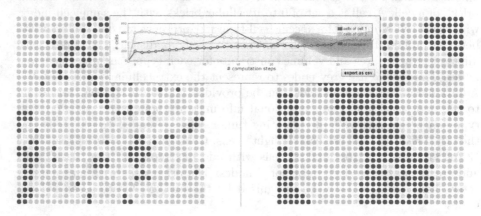

Fig. 6. Mock-Up of a time-line extension showing results of a computation taking additional information on treatment into account

analysis. Regarding the carcigonesis we have to include information about several attributes of tumor progression as well as genetic theory. Genomic databases also provide data in biotab format that includes temporal data such as "days to death" [50]. The possibilities are numerous. Comparing progress is possible with visualization metaphors such as making use of a Layer Area graph, Braided graph, Stream-graph or even parallel coordinates as well as many others [51].

5   Challenges

Our work is an intermediate step in extending cancer research using a specific tool and feeding it with additionally enhanced data. A number of challenges has to be addressed. There are many open issues for data integration, in particular to cancer data. We summarize and explain the most important ones.

Relevance. A key challenge is finding suitable relations in a domain-specific manner. Are relevant data such as growth rates explicitly available via open data sources or hidden within text retrieval of open access publications (literature curation)? How can relevant data sets be successfully retrieved?

Data Quality. Regarding data quality, aspects of accuracy and completeness have to be taken into account. Several genomics databases show associations between diseases and genes for several reasons, sometimes only because of the fact that queried terms occurred in the same publication. Further data processing steps have to be taken into account to decrease retrieval of false-positive or false-negative associations.

Tight Integration of Visualization. Integration for visual data analysis is possible on different levels. Moving beyond visualization as simple presentation of computation results, several interaction possibilities have to be included seamlessly to foster understanding of the underlying processes [5].

Specifically in the case of simulations, experts need to set many parameters but it is often not clear what the effect of the different parameters will be. Hence, there is a need for representing sensitivity and also, uncertainty of the analysis results. The latter is particularly relevant in case of incomplete data, or data of varying levels of resolution. Moreover, the integration of the knowledge of a domain expert can sometimes be indispensable, and the interaction of a domain expert with the data would greatly enhance the whole knowledge discovery process pipeline, i.e. interactive machine learning puts an human-into-the-loop to enable what neither a human nor a computer could do on their own [52].

Ease of Use. Incorporating a human computer interaction perspective into cancer simulation and visual analysis, we have to face the danger of user interface overload due to the complexity of data integration. Integrating various multi-dimensional result-sets of different databases in a consistent and concise way to maintain an intuitive user interface. While our approach is to provide tumor growth simulation and visual analysis via an intuitive user interface that is online available, questions to be answered still remain: How to facilitate exploration and discovery and how to make complex cancer data easily accessible.

6   Discussion and Conclusion

Cancer research is a data-intensive application domain that, on the one hand, raises many challenges for researchers, technicians and clinicians. On the other one in silico modeling may benefit from the many possibilities that come with accessible data related to the disease of cancer.

We implemented an easily extendable user interface using open-source components, with the ultimate goal of supporting in silico modeling by dissemination and contribution throughout the Computational Biology community for cancer research. Visualization for scientific simulations can have a positive impact on exploration, comparison and understanding. Therefore we are iteratively extending a visualization approach to tumor growth simulation and describe some examples as a starting point, how publicly available data can be used to further enhance the analysis of tumor kinetics.

We believe that it is essential to exploit and integrate data to achieve the goal of supporting clincians' decision making. The tool's extensions have been co-designed and validated by a domain-expert, but have not been evaluated by clinicians so far. Future plans are to conduct iterative testing and validating.

This contribution is preliminary work and aims to facilitate integration of heterogeneous data sources for tumor simulation and analysis by providing a categorized list of databases and describing integration possibilities. Open Data for cancer research can be disposed on a large scale: Incidence reports can be used to enhance a statistical and probabilistic approach to prediction regarding population data such as age, sex, etc. Imaging archives can be exploited for input testing. Further, profiles can be created and utilized. First attempts are discussed in [53]. Databases provide information about mutation probabilities regarding specific cancer types. Subsequently, genomic information can be used for biomarker discovery, for targeting strategies regarding novel drugs. Moreover, the comparison of biopsies with other incidence reports may foster personalized medicine. Data can be used for parameter refinement not only for extending the set of profiles but also including more variables according multicellular structures.

In general, the sheer abundance of data, derived from multiple experiments in cancer research, asks for a more comprehensive approach to data retrieval, analysis and application [36].

The progress of sophisticated biochemical and biomedical methods may not outrank the development of bioinformatic methods in order to salvage the often multi-dimensional information There is a general need to readily access cancer data from public repositories. Data integration resembles one promising option to this task.

So far, Web repositories on cancer information focus genomic and mutational data in particular. We experienced that one can easily get sunk within this magnitude of information in search of completely different readings. We aim to pick and choose details of growth-relevance in order to refine and improve kinetic models within field of computational biology in cancer. In anticipation of future development, in terms of personalized medicine, individual mutational profiles could be compared to those from repositories and integrated by determining the scope of the specific tumor growth. This approach could be equally employed for proteomic material. For that matter, further information on spatial and temporal changes due to genetic changes have to be allocated to online repositories. Ultimately, such an approach will predict the outcome of the disease and the patient's survival possibilities.

Concluding, we believe that the key to understanding the concept of cancer lies within the integrative translation and multi-dimensional connection of open data.

References

1. Holzinger, A., Dehmer, M., Jurisica, I.: Knowledge discovery and interactive data mining in bioinformatics - State-of-the-Art, future challenges and research directions. BMC Bioinform. **15**(Suppl. 6), I1 (2014)
2. Jeanquartier, F., Jean-Quartier, C., Cemernek, D., Holzinger, A.: In silico modeling for tumor growth visualization. BMC Syst. Biol. (2016)
3. Hey, T., Tansley, S., Tolle, K.: The Fourth Paradigm: Data-Intensive Scientific Discovery. Microsoft Research (2009)
4. Ward, M.O., Grinstein, G., Keim, D.: Interactive Data Visualization: Foundations, Techniques, and Applications. CRC Press, Natick (2010)
5. Turkay, C., Jeanquartier, F., Holzinger, A., Hauser, H.: On computationally-enhanced visual analysis of heterogeneous data and its application in biomedical informatics. In: Holzinger, A., Jurisica, I. (eds.) Knowledge Discovery and Data Mining. LNCS, vol. 8401, pp. 117–140. Springer, Heidelberg (2014)
6. Unger, A., Schumann, H.: Visual support for the understanding of simulation processes. In: IEEE Pacific Visualization Symposium, PacificVis 2009, pp. 57–64. IEEE (2009)
7. Bernard, J., Daberkow, D., Fellner, D., Fischer, K., Koepler, O., Kohlhammer, J., Runnwerth, M., Ruppert, T., Schreck, T., Sens, I.: VisInfo: a digital library system for time series research data based on exploratory search - a user-centered design approach. Int. J. Digit. Libr. **1**, 37–59 (2015). Springer
8. Bernard, J., Ruppert, T., Scherer, M., Kohlhammer, J., Schreck, T.: Content-based layouts for exploratory metadata search in scientific research data. In: Proceedings of the 12th ACM/IEEE-CS Joint Conference on Digital Libraries, pp. 139–148. ACM, June 2012
9. Scherer, M., von Landesberger, T., Schreck, T.: Visual-interactive querying for multivariate research data repositories using bag-of-words. In: Proceedings of ACM/IEEE Joint Conference on Digital Libraries, pp. 285–294 (2013)
10. Shao, L., Behrisch, M., Schreck, T., von Landesberger, T., Scherer, M., Bremm, S., Keim, D.: Guided sketching for visual search and exploration in large scatter plot spaces. In: Proceedings of EuroVA International Workshop on Visual Analytics, pp. 19–23 (2014)
11. Kandel, S., Paepcke, A., Hellerstein, J., Wrangler, J.H.: Interactive visual specification of data transformation scripts. In: ACM Human Factors in Computing Systems (CHI) (2011)
12. Jeanquartier, F., Jean-Quartier, C., Holzinger, A.: Integrated Web visualizations for protein-protein Interaction databases. BMC Bioinform. **16**(1), 195 (2015). doi:10.1186/s12859-015-0615-z
13. Gomez-Cabrero, D., Abugessaisa, I., Maier, D., Teschendorff, A., Merkenschlager, M., Gisel, A., Tegnér, J.: Data integration in the era of omics: current and future challenges. BMC Syst. Biol. **8**(Suppl. 2), I1 (2014)
14. Angrist, M., Cook-Deegan, R.: Distributing the future: the weak justifications for keeping human genomic databases secret and the challenges and opportunities in reverse engineering them. Appl. Transl. Genomics **3**(4), 124–127 (2014)

15. Cerami, E., Gao, J., Dogrusoz, U., Gross, B.E., Sumer, S.O., Aksoy, B.A., Antipin, Y.: The cBio cancer genomics portal: an open platform for exploring multidimensional cancer genomics data. Cancer Discov. **2**(5), 401–404 (2012)

16. Cline, M.S., Craft, B., Swatloski, T., Goldman, M., Ma, S., Haussler, D., Zhu, J.: Exploring TCGA pan-cancer data at the UCSC cancer genomics browser. Sci. Rep. **3**, 2652 (2013)

17. Beroukhim, R., Mermel, C.H., Porter, D., Wei, G., Raychaudhuri, S., Donovan, J., Mc Henry, K.T.: The landscape of somatic copy-number alteration across human cancers. Nature **463**(7283), 899–905 (2010)

18. Forbes, S.A., Beare, D., Gunasekaran, P., Leung, K., Bindal, N., Boutselakis, H., Kok, C.Y.: COSMIC: exploring the world's knowledge of somatic mutations in human cancer. Nucleic Acids Res. **43**(D1), D805–D811 (2015)

19. Zhang, J., Baran, J., Cros, A., Guberman, J.M., Haider, S., Hsu, J., Wong-Erasmus, M.: International Cancer Genome Consortium Data Portala one-stop shop for cancer genomics data. Database (Oxford) (2011) bar026

20. Rubio-Perez, C., Tamborero, D., Schroeder, M.P., Antoln, A.A., Deu-Pons, J., Perez-Llamas, C., Lopez-Bigas, N.: In silico prescription of anticancer drugs to cohorts of 28 tumor types reveals targeting opportunities. Cancer Cell **27**(3), 382–396 (2015)

21. Thorvaldsdttir, H., Robinson, J.T., Mesirov, J.P.: Integrative Genomics Viewer (IGV): high-performance genomics data visualization and exploration. Briefings Bioinform. **14**(2), 178–192 (2013)

22. Dietmann, S., Lee, W., Wong, P., Rodchenkov, I., Antonov, A.V.: CCancer: a birds eye view on gene lists reported in cancer-related studies. Nucleic Acids Res. **38**(Suppl. 2), W118–W123 (2010)

23. Jiang, G., Sohn, S., Zimmermann, M.T., Wang, C., Liu, H., Chute, C.G.: Drug normalization for cancer therapeutic and druggable genome target discovery. AMIA Summits Transl. Sci. Proc. **2015**, 72 (2015)

24. Clark, K., Vendt, B., Smith, K., Freymann, J., Kirby, J., Koppel, P., Moore, S., Phillips, S., Maffitt, D., Pringle, M., Tarbox, L., Prior, F.: The Cancer Imaging Archive (TCIA): maintaining and operating a public information repository. J. Digit. Imaging **26**(6), 1045–1057 (2013)

25. Ongenaert, M., Van Neste, L., De Meyer, T., Menschaert, G., Bekaert, S., Van Criekinge, W.: PubMeth: a cancer methylation database combining text mining and expert annotation. Nucleic Acids Res. **36**(Suppl. 1), D842–D846 (2008)

26. Zhu, F., Patumcharoenpol, P., Zhang, C., Yang, Y., Chan, J., Meechai, A., Shen, B.: Biomedical text mining and its applications in cancer research. J. Biomed. Inform. **46**(2), 200–211 (2013)

27. Pletscher-Frankild, S., Pallej, A., Tsafou, K., Binder, J.X., Jensen, L.J.: DISEASES: text mining and data integration of diseasegene associations. Methods **74**, 83–89 (2015)

28. Holzinger, A., Schantl, J., Schroettner, M., Seifert, C., Verspoor, K.: Biomedical text mining: state-of-the-art, open problems and future challenges. In: Holzinger, A., Jurisica, I. (eds.) Knowledge Discovery and Data Mining. LNCS, vol. 8401, pp. 271–300. Springer, Heidelberg (2014)

29. Torre, L.A., Siegel, R.L., Ward, E.M., Jemal, A.: Global cancer incidence and mortality rates and trendsan update. Cancer Epidemiol. Biomark. Prev. **25**(1), 16–27 (2016)

30. Siegel, R.L., Miller, K.D., Jemal, A.: Cancer statistics, 2016. CA: A Cancer J. Clin. **66**(1), 7–30 (2015)

31. Bray, F., Ferlay, J., Laversanne, M., Brewster, D.H., Gombe Mbalawa, C., Kohler, B., Soerjomataram, I.: Cancer incidence in five continents: inclusion criteria, highlights from Volume X and the global status of cancer registration. Int. J. Cancer **137**(9), 2060–2071 (2015)
32. Europe PMC Consortium: Europe PMC: a full-text literature database for the life sciences and platform for innovation. Nucleic Acids Res. **43**(D1), D1042–D1048 (2015)
33. Holzinger, A., Jurisica, I.: Knowledge discovery and data mining in biomedical informatics: the future is in integrative, interactive machine learning solutions. In: Holzinger, A., Jurisica, I. (eds.) Knowledge Discovery and Data Mining. LNCS, vol. 8401, pp. 1–18. Springer, Heidelberg (2014)
34. Kieseberg, P., Weippl, E., Holzinger, A.: Trust for the doctor-in-the-loop. In: European Research Consortium for Informatics and Mathematics (ERCIM) News: Tackling Big Data in the Life Sciences, vol. 104(1), pp. 32–33 (2016)
35. Greiling, D.A., Jacquez, G.M., Kaufmann, A.M., Rommel, R.G.: Space-time visualization and analysis in the Cancer Atlas Viewer. J. Geogr. Syst. **7**(1), 67–84 (2005)
36. Wei, Y.: Integrative analyses of cancer data: a review from a statistical perspective. Cancer Inform. **14**(Suppl. 2), 173 (2015)
37. Wu, T.J., Schriml, L.M., Chen, Q.R., Colbert, M., Crichton, D.J., Finney, R., Mitraka, E.: Generating a focused view of disease ontology cancer terms for pan-cancer data integration and analysis. Database (2015) bav032
38. Sioutos, N., de Coronado, S., Haber, M.W., Hartel, F.W., Shaiu, W.L., Wright, L.W.: NCI Thesaurus: a semantic model integrating cancer-related clinical and molecular information. J. Biomed. Inform. **40**(1), 30–43 (2007)
39. Drake, J.W., Charlesworth, B., Charlesworth, D., Crow, J.F.: Rates of spontaneous mutation. Genetics **148**(4), 1667–1686 (1998)
40. Lodish, H., Berk, A., Zipursky, S.L., et al.: Molecular Cell Biology, 4th edn. W.H. Freeman, New York (2000)
41. Yang, Y., Dong, X., Xie, B., Ding, N., Chen, J., Li, Y., Fang, X.: Databases and web tools for cancer genomics study. Genomics Proteomics Bioinform. **13**(1), 46–50 (2015)
42. Müller, H.M., Kenny, E.E., Sternberg, P.W.: Textpresso: an ontology-based information retrieval and extraction system for biological literature. PLoS Biol. **2**(11), e309 (2004)
43. Schaefer, C., Grouse, L., Buetow, K., Strausberg, R.L.: A new cancer genome anatomy project web resource for the community. Cancer J. **7**(1), 52–60 (2001)
44. Bult, C.J., Krupke, D.M., Begley, D.A., Richardson, J.E., Neuhauser, S.B., Sundberg, J.P., Eppig, J.T.: Mouse Tumor Biology (MTB): a database of mouse models for human cancer. Nucleic Acids Res. **43**(D1), D818–D824 (2015)
45. Roelofs, E., Dekker, A., Meldolesi, E., van Stiphout, R.G., Valentini, V., Lambin, P.: International data-sharing for radiotherapy research: an open-source based infrastructure for multicentric clinical data mining. Radiother. Oncol. **110**(2), 370–374 (2014)
46. WHO cancer mortality database (IARC). http://www-dep.iarc.fr/WHOdb/WHOdb.htm. Accessed 01 May 2016
47. Eyler, C.E., et al.: Glioma stem cell proliferation and tumor growth are promoted by nitric oxide synthase-2. Cell **146**(1), 53–66 (2011)
48. Herman, A.B., Savage, V.M., West, G.B.: A quantitative theory of solid tumor growth, metabolic rate and vascularization. PLOS One **6**, e22973 (2011)

49. Kisker, O., Becker, C.M., Prox, D., Fannon, M., D'Amato, R., Flynn, E., Fogler, W.E., Kim Lee Sim, B., Allred, E.N., Pirie-Shepherd, S.R., Folkman, J.: Continuous administration of endostatin by intraperitoneally implanted osmotic pump improves the efficacy and potency of therapy in a mouse xenograft tumor model. Cancer Res. **61**, 7669 (2001)

50. Mroz, E.A., Tward, A.M., Hammon, R.J., Ren, Y., Rocco, J.W.: Intra-tumor genetic heterogeneity and mortality in head and neck cancer: analysis of data from the cancer genome atlas. PLoS Med. **12**(2), e1001786 (2015)

51. Aigner, W., Miksch, S., Schumann, H., Tominski, C.: Visualization of Time-oriented Data. Springer Science & Business Media, New York (2011)

52. Holzinger, A.: Interactive machine learning for health informatics: when do we need the human-in-the-loop? Brain Inform. **3**(2), 119–131 (2016). Springer

53. Jean-Quartier, C., Jeanquartier, F., Cemernek, D., Holzinger, A.: Tumor growth simulation profiling. In: Renda, M.E., Bursa, M., Holzinger, A., Khuri, S. (eds.) ITBAM 2016. LNCS, vol. 9832, pp. 208–213. Springer, Heidelberg (2016)

Information Technologies in Brain Science

Filter Bank Common Spatio-Spectral Patterns for Motor Imagery Classification

Ayhan Yuksel[(✉)] and Tamer Olmez

Department of Electronics and Communication Engineering,
Istanbul Technical University, Istanbul, Turkey
yukselay@itu.edu.tr

Abstract. In this study, a new spatio-spectral filtering method for motor imagery signal analysis is introduced. Motor imagery is an important research area in brain computer interfacing. EEG signals related with motor imagery have characteristic frequencies originating from sensorimotor cortex. Common spatial patterns (CSP) method is a very popular and successful spatial filtering algorithm in motor imagery classification. However, CSP only optimizes spatial filters, subject specific frequency selection should be done manually, which is a meticulous process. Therefore, an automatic method for spectral filter optimization is needed. Proposed filter bank common spatio-spectral patterns (FBCSSP) algorithm optimizes spatial and spectral filters. FBCSSP method uses a network of a filter bank and two consecutive CSP layers so that proposed structure has a subject specific response in both spatial and spectral domains. We inspected the proposed method in terms of classification accuracy and physiological consistence of the created filters using publicly available data set. FBCSSP method gave higher classification accuracy than other spatio-spectral pattern methods in the literature. Also, obtained spatial and spectral filters were consistent with the spatial and spectral properties of motor imagery signals.

Keywords: Brain computer interfaces (BCI) · Motor imagery (MI) · Electroencephalogram (EEG) · Common spatial patterns (CSP)

1 Introduction

Brain computer interface (BCI) is an assistive technology which helps disabled people by setting up a direct communication link between the users brain and an electronic device or software such as a wheel chair, a computer running a word processing program or a quad-copter [10,18]. In a BCI system, the activity of the brain is measured and then converted to the control commands for the controlled device. There are many techniques for measuring the brain activity such as functional magnetic resonance imaging (FMRI), magnetoencephalography (MEG), positron emission tomography (PET), electrocorticogram (ECoG) or electroencephalography (EEG). Among them, EEG is the preferred way of

© Springer International Publishing Switzerland 2016
M.E. Renda et al. (Eds.): ITBAM 2016, LNCS 9832, pp. 69–84, 2016.
DOI: 10.1007/978-3-319-43949-5_5

acquiring brain signals thanks to its practicality, being low cost, non-invasiveness and portability [20].

Motor imagery (MI) is an independent BCI method which uses motor cortex as a signal source. MI guesses motor intentions of the user without any actual muscular movement. In this context, the user imagines moving a limb while his/her EEG is analyzed continuously by a BCI system which finds out the imagined movement. Then, a command to be sent to the controlled device is generated according to the type of the imagined motor movement. Motor imagery studies showed that, imagination of movement of a limb creates special oscillations called event related synchronization (ERS) and event related de-synchronization (ERD) at specific frequency bands [2,15].

The pioneer study of Penfield and Boldrey in 1937 revealed important information about spatial organization of motor cortex. They reported that, any muscle group in the body is presented at a specific area on the motor cortex. The sizes of these areas are proportional with the usage skills of the corresponding limb rather than the limbs real size. The popular figure named homunculus resizes the limbs proportional to their areas occupied on the motor cortex. The spatial organization on motor cortex yields identification of various motor imagery tasks according to the location of event related synchronization (ERS) and de-synchronization (ERD) rhythms. However, scalp EEG signal is seriously affected by the volume conduction effect in which the EEG signals all over the scalp are mixed up and this results in poor spatial resolution [5]. In order to remove the volume conduction effect, some spatial filtering methods are proposed such as common average reference (CAR) [12], Laplacian (LAP) [12], common spatial filters (CSP) [16] and spatial filter network (SFN) [19]. Among these methods, CSP is a well known method for motor imagery classification problem and it was proven to be efficient in recent BCI competitions [3,4].

To focus on the event related synchronization and de-synchronization (ERD and ERS) signals and to achieve a high classification performance, it is necessary to filter the EEG signal with a band pass filter prior to CSP calculation. However, one problem is that, the frequency bands of these signals vary from subject to subject. Generally, the cut-off frequencies of the band pass filter are either selected manually or unspecifically set to a broad band filter [7], which results in poor classification performance. Manual searching of the best frequency band through the training set is laborious and time consuming [13]. Thus, optimizing a spectral filter along with the spatial filter is highly desirable [7].

Common spatio-spectral pattern (CSSP) algorithm [11] is the firstly proposed method to address this problem. CSSP embeds time delayed channels into the original EEG signal in order to create a first order FIR filter for each channel. Obtained results showed an improvement of the CSSP algorithm over CSP. However, a first order FIR filter is very limited to select a certain frequency band from the EEG spectrum. After that, an improvement to CSSP, Common sparse spectral spatial pattern (CSSSP) was proposed [7]. CSSSP designs a FIR filter with any order and common to all channels. This method searches for a set of spectral-spatial filter coefficients by gradient search method which is computationally expensive with additional cost for sparsification and it needs some parameter tunings.

Sub-band common spatial patterns (SBCSP) [13] and filter bank common spatial patterns (FBCSP) [1] methods are based on optimizing spatial filters for multiple spectral filters that have different pass-bands. As reported in BCI Competition III and IV, FBCSP method achieved a high classification accuracy [4]. In these methods, a filter bank is used in order to decompose EEG signal into multiple frequency bands and a separate spatial filter is calculated for each band by CSP method. Then, features belong to different frequency bands are chosen by feature selection methods based on mutual information maximization.

Higashi et al. recently proposed a method for simultaneous design of spectral and spatial filters [9] called discriminative filter bank CSP (DFBCSP). DFBCSP algorithm optimizes the coefficients of FIR filter(s) and corresponding spatial weights. DFBCSP proposes an iterative method to optimize the spatial and spectral filter coefficients by converting the spatial and spectral optimization problems into separate generalized eigen value problems. Since it is an iterative method, reaching the optimum point should take many steps and optimization speed of the DFBCSP method depends on the degree of the FIR filter to be optimized.

In this paper, we present spatio-spectral filtering method which binds the spatial and the spectral filters in a mixed architecture that we call filter bank common spatio - spectral patterns (FBCSSP). FBCSSP finds out the required filter parameters with simple CSP calculations in one pass, without any iteration. The detailed description of FBCSSP is found in Materials and Methods section. We then compare the proposed method with other spatio spectral filtering methods in the literature and obtained results shows higher classification accuracy over them.

This paper is organized as follows, in Sect. 2, CSP and the proposed FBCSSP method will be described in details. In Sect. 3 we give the evaluations of the proposed method comparing with other methods. Section 4 investigates the advantages and the disadvantages of the FBCSSP algorithm. Finally, the Conclusion section summarizes the study and concludes the paper.

2 Materials and Methods

2.1 Common Spatial Patterns

Let X_k to be the kth epoch with class c in a motor imagery experiment which includes N EEG channels and T time samples that are filtered with a band bass filter which is manually set at a fixed frequency band. Let $w \in \mathbb{R}^{Nx1}$ to be an N dimensional spatial filter. Spatial filtering is simply the linear combination of the channels with the coefficients of w:

$$z_k = w^\top X_k \tag{1}$$

where $z_k \in \mathbb{R}^{1xT}$ denotes the projection of epoch X_k and $^\top$ is the transpose operation. CSP method searches for the best filter which maximizes the average power of one class while minimizing the average power of the other class. Since the epoch X_k is a zero average signal ($\mu_k = 0$) as a result of band pass filtering, power of z_k is obtained by the variance calculation:

$$P_k^\mathsf{T} = \sigma^2\left(z_k\right) = \frac{1}{T}\sum_{t=1}^{T}\left|w^\mathsf{T}\left(X_k(t) - \mu_k\right)\right|^2 = w^\mathsf{T} R_k w \qquad (2)$$

Where $R_k \in \mathbb{R}^{NxN}$ is the covariance matrix of epoch k . Let $R_c \in \mathbb{R}^{NxN}$ to be the average covariance matrices of the epochs that belong to the class c:

$$R^{(c)} = \frac{1}{n_c}\sum_{k\in c}^{n_c} R_k \qquad (3)$$

CSP uses Rayleigh ratio as an optimization function, which is the ratio of average powers after spatial filtering:

$$w_{csp} = \arg\max_{w}\frac{w^\mathsf{T} R^{(1)} w}{w^\mathsf{T} R^{(2)} w} \qquad (4)$$

This optimization problem is solved by converting it to a generalized eigen value problem:

$$(R^{(2)^{-1}} R^{(1)})w = \lambda w \qquad (5)$$

Since the covariance matrices are of dimension NxN, solution to the generalized eigen value problem above generates n eigen vector (w) - eigen value (λ_n) pairs ($n = 1, 2, ...N$). Note that, for any solution w_n, Rayleigh ratio in (4) gives λ_n. Thus, eigen vector corresponding to the largest eigen value gives the maximum power ratio for class 1 over class 2 and, eigen vector corresponding to the smallest eigen value gives the maximum power ratio for class 2 over class 1. So, CSP firstly sorts the eigen values in descending order:

$$\lambda_1 > \lambda_2 > \cdots > \lambda_m > \lambda_{m+1} > \cdots\cdots > \lambda_{N-m} > \lambda_{N-m+1} > \cdots \lambda_N \qquad (6)$$

and then gets the m upper and m lower eigen vectors in order to create a spatial filter matrix $W \in \mathbb{R}^{MxN}$, where $M = 2m$. To classify an input epoch with unknown class label, CSP firstly creates feature vectors using epochs with known class labels. Feature vector of an epoch is usually log-variance of the spatially filtered signal:

$$f_j = log\left(\frac{var(z^j)}{\sum_{l=1}^{M} var(z^l)}\right) \quad j = 1, 2, ...M \qquad (7)$$

where j represents the column number in feature vector $f \in \mathbb{R}^M$. In the above equation, logarithm function is used for approximating the feature distribution to a normal distribution [8].

2.2   Filter Bank Common Spatio - Spectral Patterns

The proposed method of this study called filter bank common spatio - spectral patterns (FBCSSP) method consists of two CSP layers. At the first layer, EEG signal is filtered with a couple of FIR band pass filters. Then, each band passed EEG signal with N channels are spatially filtered in the CSP-1 layer so that, best spatial patterns for each frequency bands are determined in this layer. At this point, proposed method differs from FBCSP and SBCSP methods. These methods finalize preprocessing and extract features at the end of the first layer. However, FBCSSP method continues signal preprocessing operation. Obtained CSP-1 outputs are directly given to a second CSP filter, CSP-2. The purpose of CSP-2 is linearly combining the outputs of the first spatial filter layer so that maximum divergence could be obtained.

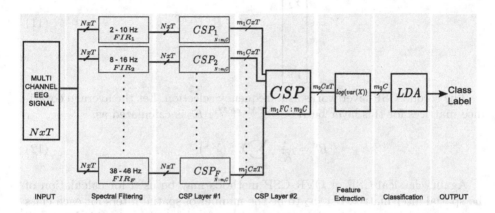

Fig. 1. A flowchart regarding filter bank common spatio - spectral patterns method.

Let $X \in \mathbb{R}^{NxT}$ be an input EEG signal matrix with T samples and N channels, called as epoch. Firstly, all epochs in the training set are filtered with FIR band pass filters at desired frequencies with degree P,

$$\hat{X}_{f,n}(t) = \sum_{p=0}^{P} h_{f,p} X_n(t-p) \qquad \hat{X}_f \in \mathbb{R}^{NxT} \quad f = (1, 2, \cdots F) \qquad (8)$$

Where, $h_{f,p}$ is the pth weight of fth FIR filter. For each FIR filter output, a CSP filter is created. Let the average covariance matrices at the output of the fth filter be R_f^c where, c is the class label,

$$R_f^c = \frac{1}{K_c} \sum_{k \in c} \hat{X}_f^k \left(\hat{X}_f^k \right)^{\top} \qquad (9)$$

Where K_c is the number of epochs which belong to the class c. For two classes, classical CSP approach may be applied. However, in order to find the spatial filters in a multiclass BCI experiment, one versus rest (OVR) CSP method may be applied [6]. Let m_1 o denote the number of spatial filters for one class at the first layer and M_1 to denote the total number of spatial filters where $M_1 = m_1 C$ and C is the total number of classes. Let the obtained spatial filter to be denoted with $U_f \in \mathbb{R}^{N x M_1}$. Then, the output of this spatial filter will be,

$$Y_f = U_f^\top \hat{X}_f \qquad Y_f \in \mathbb{R}^{M_1 x T} \tag{10}$$

For the next layer, all outputs of first spatial layer are concatenated row by row and a new epoch is created. Let Y be the new epoch matrix which is defined as,

$$Y = \begin{bmatrix} Y_1 \\ Y_2 \\ \vdots \\ Y_F \end{bmatrix} \qquad Y \in \mathbb{R}^{F M_1 x T} \tag{11}$$

The second CSP layer works as a frequency selection. Let the average covariance matrices for this layer be $R^c \in \mathbb{R}^{F M_1 x F M_1}$. R^c is calculated as,

$$R^c = \frac{1}{K_c} \sum_{k \in c} Y^k \left(Y^k \right) \top \tag{12}$$

Again, classical CSP or OVR CSP methods may be used for calculation of the spatial filter matrix. Let m_2 to be the number of spatial filters for each class at the second layer. Then, the total number of spatial filters in this layer will be $M_2 = m_2 C$. If we denote $W \in \mathbb{R}^{F M_1 x M_2}$ as the spatial filter matrix of the second label, the output of this layer will be,

$$Z = W^\top Y \qquad Z \in \mathbb{R}^{M_2 x T} \tag{13}$$

In the feature extraction method, same equation is used with CSP method that was given in (7). Note that, in this case there will be M_2 features. Obtained features is given to a linear classifier such as LDA.

2.3   Filter Bank Selection

The filter bank used in FBCSSP may be configured according to the requirements and prior information related with the processed signal. For motor imagery signals, a filter bank covering 8 Hz to 36 Hz is reasonable. However, FBCSSP has the capability to combine the output filters and finally generate an optimized spectral filter. Therefore, it is better to choose a filter bank which covers a wide frequency band in which the banks overlap.

While creating the filter bank structure, having linear phase response is the most important point because in the second CSP layer, a spectral combination operation is done. FIR filters are appropriate option for being linear phase response.

Here, the effect of linear combining the filter bank outputs will be analyzed. Embedding (8) into (10) gives the output of the first layer in terms of the input and the spectral filter parameters,

$$Y_f(t) = U_f^\top \sum_{p=0}^{P} h_{f,p} X(t-p) \tag{14}$$

By using (13), it is possible to write the overall filter within one equation,

$$Z(t) = \sum_{f=1}^{F} W_f^T U_f^\top \sum_{p=0}^{P} h_{f,p} X(t-p) \tag{15}$$

where, $W_f \in \mathbb{R}^{M_1 x M_2}$ is the spatial filter matrix in the second layer, associated with the fth output of the first layer. By organizing this equation, we reach the equation,

$$Z(t) = \sum_{p=0}^{P} \delta_p^\top X(t-p) \tag{16}$$

which is the characteristic equation of the proposed spatio - spectral filter. $\delta_p \in \mathbb{R}^{NxD}$ holds the spatial and spectral characteristics of the FBCSSP filter and it is defined as the following equation.

$$\delta_p = \sum_{f=1}^{F} U_f W_f \tag{17}$$

Above result yields, linear combination of the outputs of FIR filters means linear combination of the filter coefficients. So, the both CSP layers bring up a FIR filter which is a combination of the banks in the filter bank. Therefore, using overlapped and numerous filters should give more flexible FIR filters. An example filter bank frequency response is given in Fig. 4. Here, there are 7 FIR filters that cover the entire frequency band and their linear combination.

Normally, CSP filter extracts independent components while simultaneously diagonalizing of two covariance matrices. So, applying the CSP method to the output of another CSP filter will not improve the divergence of the signal since the output of the first CSP filter is linearly independent. However in the proposed method, when joined altogether, the outputs of the first layer will become a non-independent multi channel signal thus, CSP of second layer should increase the overall divergence of the incoming signal. In fact, second CSP makes a spectral weighting while linearly combining the outputs of the first layer.

Since they are linear filters of the same type, instead of cascading CSP filters one after another, a single CSP could be used. Indeed, this should give a higher

Rayleigh ratio then the FBCSSP. However, this CSP matrix filter would have NxF inputs and M_2 outputs and classifying performance will not be higher as expected.

Different to various spectral filter optimization methods reported in the literature, FBCSSP searches for the linear combinations of some predefined FIR filters. FBCSSP method has some advantages over these methods. Firstly, computational complexity of the algorithm will not increase with the degree of the FIR filters. Because only the outputs of the filters are being used, algorithm does not try to manipulate the filter weights. Whereas, those methods face with increasing computational complexity with the increasing filter degree. Secondly, FBCSSP method gives the flexibility of defining various FIR filters. This makes the algorithm to embed the existing prior knowledge into the spectra-spatial filters. For example, one can design FIR filters especially at the spectral region of μ and β waves and ignore the other frequencies. Third advantage of the FBCSSP is its non-iterative structure. The methods in [7,9,17] use an alternating optimization strategy which iteratively increases the fitness function by updating spatial and spectral filter parameters, respectively. Different spatial locations at different frequency bands may be activated in execution of motor imagery. This leads to spatial patterns specific to frequency band. FBCSSP method does not ignore this assumption and produces spatial filters for each defined frequency bands. This enables us to investigate the obtained spatial-spectral filters at a specific frequency band.

2.4   Data Description and Preprocessing

We used the data set from BCI competition III, which is data set IVA. Detailed information about this BCI competition may be found in [4]. Data set IVA is a 2 class motor imagery data set which includes EEG records from 5 subjects labeled AA, AL, AV, AW and AY. Actually there were three classes ('right', 'left' and 'foot') in the original experiment, only cues for the classes 'right' and 'foot' are provided in the public dataset. The recording includes 118 EEG channels were measured at positions of the extended international 10/20-system. Signals were sampled at 1000 Hz digitized with 16 bit (0.1uV) accuracy and band-pass filtered between 0.05 and 200 Hz. Classes were labeled as right hand and foot. For each subject, there are 280 trials defined with starting and ending markers as well as its class label.

In this study, we applied the same pre-processing steps to all subjects. (i) we selected electrodes on the motor cortex area. Selected electrodes for the two data sets are shown in Fig. 2. (ii) For CSP method, EEG is band pass filtered with 8–30 Hz 5th order Butterworth filter since this band covers the motor imagery signal frequency range roughly. For spatio spectral filtering methods, we used a 5 th order Butterworth filter with a pass band of 1–49 Hz. (iii) For each trial, we used EEG signals in time segment between 0.5 s–2.5 s after instruction cue. Also trials marked with rejected trial were excluded. Preprocessing phase is given in Fig. 3.

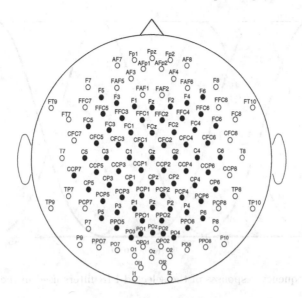

Fig. 2. EEG channels used in BCI competition III Data Set IVa. EEG was captured with 118 electrodes according to the extended international 10/20-system

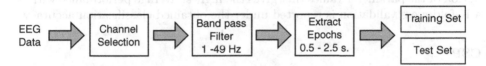

Fig. 3. Preprocessing progress used in the evaluation

2.5 Selected filters

In this study, we used a total of seven overlapping band pass FIR filters which cover the frequency band 1 50 Hz. The degree of the filters (P) was set to 20. Note that one can search for different filter bank configuration providing that only FIR filters are used because of their linear phase response. Since the selected filters overlap, their linear combination should produce new spectral filter specific to the subject under test. Frequency response of the filters used in the study is given in Fig. 4.

3 Results

3.1 Evaluated Methods and Selected Configurations

In the following paragraphs, the methods that were evaluated will be listed with the configuration specific to the method itself. Presented classification algorithms have at least one setting values that is called *hyper-parameter*. For each subjects

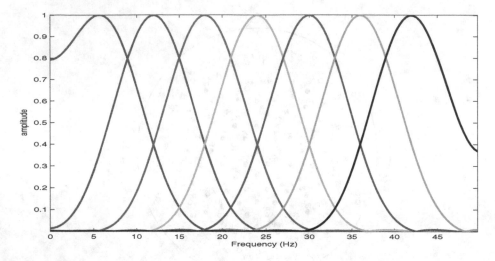

Fig. 4. Frequency responses of the selected FIR filters used in the study

and each method, hyper parameters are selected from a list and the best combination of the parameter values that gives the highest average performance within a K-fold cross validation is reported under the obtained classification accuracy.

CSP

CSP method uses m as a hyper parameter which stands for the number of spatial filters per class used for constructing the spatial filter. Possible values for m was selected from $\{1, 2, 3, 4, 5\}$. Then for each subject, those possible m values were tried was tried one by one and the best m that gives the highest average performance was selected.

FBCSP

For FBCSP method, the filter bank used was a FIR filter bank including 7 FIR filters with cut off frequencies [2,10; 8,16;14,22; 20,28;26,34;32,40;38,46]. The degree of the filters was set to 20. The frequency responses of the filters in the filter bank configured for evaluation are given in Fig. 4. The parameter m for FBCSP method was chosen out of $\{1, 2, 3, 4, 5\}$ and the number of features selected (d) was chosen out of $\{1, 2, 3 \cdots 10\}$.

FBCSSP

In FBCSSP method, the hyper parameters that were tried for the best combination are $\{m_1, m_2\}$ which are the number of spatial filters for the first and the second CSP layers per class, respectively. the values of m_1 and m_2 are selected from the list $\{1, 2, 3, 4, 5\}$ so that there are 25 combinations for each subject.

The FIR filter bank used for FBCSSP method includes 7 FIR filters with cut off frequencies [2,10; 8,16;14,22; 20,28;26,34;32,40;38,46]. The degree of the filters was set to 20. The frequency responses of the filters in the filter bank configured for evaluation are given in Fig. 4.

3.2   Classification Results

The classification accuracies of the described methods for the two BCI competition data sets are listed in Table 1. The outputs of the methods listed here were classified using standard LDA classifier. Ten fold cross validation classification accuracy was calculated for all methods. It is obvious that FBCSSP performs high classification accuracy. Short description and specific configurations for each of the methods used were given in the previous sub section.

Table 1 reports classification accuracies of each method for each subject. The formula of the percentage accuracy ($ACC\%$) was given by the formula,

$$ACC\% = 100 \sum_{c=1}^{C} \frac{TP_c}{TP_c + TN_c + FP_c + FN_c} \tag{18}$$

Where, C is the total number of classes in the data set. The table also lists the standard deviation (std) of any method for any subject. Since 10-fold cross validation was used for evaluating the classification accuracy, std. represents the standard deviation of all folds for the given subject and method.

Table 1. Classification performances of the listed methods for the subjects in BCI competition III Data Set IVa.

METHOD		SUBJECTS					Average
		aa	al	av	aw	ay	
CSP	ACC (%)	83.57	97.50	72.85	92.5	94.28	88.14
	std.	±9.85	±3.39	±10.13	±7.23	±3.84	±9.99
	(*m*)	(2)	(3)	(5)	(2)	(4)	
FBCSP	ACC (%)	90.46	98.54	67.43	97.72	96.07	90.05
	std.	±4.31	±2.23	±6.63	±2.47	±3.63	±13.03
	(*m,d*)	(2;4)	(2;3)	(5;10)	(3;8)	(1;4)	
FBCSSP	ACC (%)	92.51	98.93	80.72	97.51	97.51	93.43
	std.	±3.12	±1.72	±9.55	±2.41	±2.41	±7.51
	(*m₁,m₂*)	(1;1)	(1;1)	(4;4)	(5;3)	(1;5)	

The last column lists the overall accuracy and standard deviation for any method and all subjects which summarizes the corresponding method's classification performance.

The number(s) in parenthesizes in any cell notifies the selected values of parameters for the corresponding subject and the method. Also, the names of the parameters are given in the second column. Note that, the accuracy value in each cell is the outcome of the given parameter configuration.

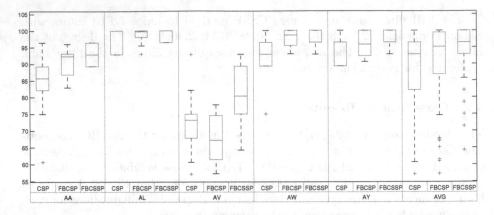

Fig. 5. Boxplots displaying the disturbance of classification performances for the subjects in dataset IVA. (Color figure online)

Since the supplied data is subjective, methods' performances highly vary along the subjects. Therefore, along with the quantitative performance summary supplied by the tables, graphical presentations of the performances which benchmark the listed methods with box plots subject by subject are given in Fig. 5. The figure shows the classification accuracies of the subjects belong to the data set IVA. In this figure, the horizontal axis is the methods that were evaluated and the vertical axis is the evaluated performance value. For any subject and any method, the box boundaries represent the upper and lower 25 % quartiles of the input data which is the output of the 10 fold cross validation for the selected configuration. The red horizontal lines inside or on the boundary of the boxes represent the median values. The whiskers (dashed lines above and below the boxes) extend to the most extreme data points the box plot algorithm considers to be not outliers, and the outliers are plotted with red cross marks individually.

3.3 Spatial and Spectral Filters

Spatial filters calculated by training CSP and FBCSSP are given in Fig. 6 for all of the subjects in the BCI data set. Also, spectral filters of FBCSSP are given in Fig. 7. Frequency response of the trained FBCSSP network is calculated by scanning all of the inputs with signal at a given frequency and measuring the average power at the output. In the figures, the spectral filter response is normalized. Spatial filter illustrations are prepared similarly, all network inputs are scanned by inputting an impulse and measuring the power of the signal for each input at the output. Then, the calculated power corresponding to any input is converted to a gray scale color value and displayed on a head figure with electrodes located. The pass band of the obtained spectral filters are located approximately within the band, which is associated with the sensorimotor cortex [14]. Besides, most of the spatial filters successfully focused on the area related with the corresponding motor action over the

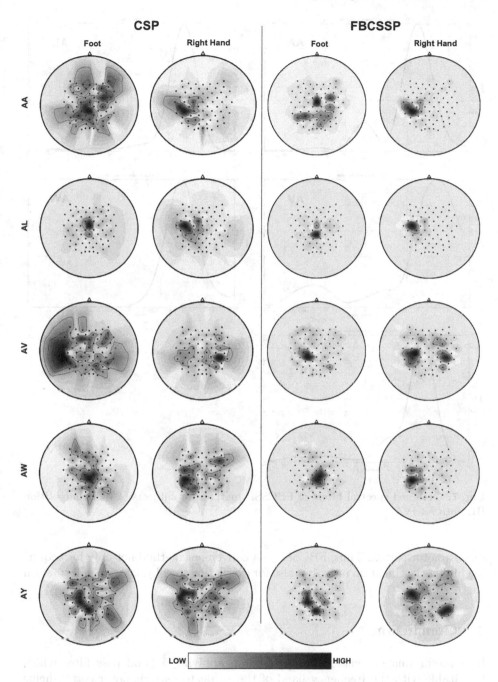

Fig. 6. Obtained spatial filters of CSP (Left) and FBCSSP (Right) methods for subjects of BCI competition III, data set IVA

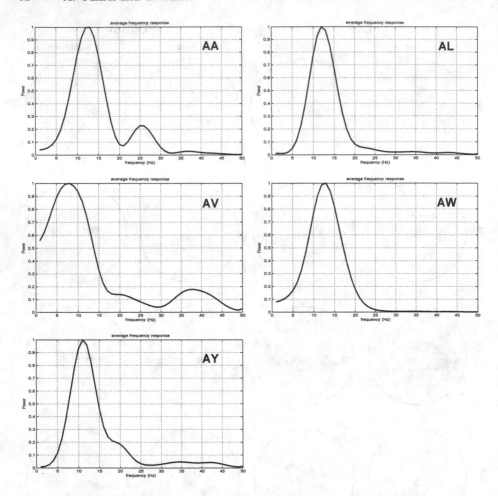

Fig. 7. Obtained spectral filters of FBCSSP method for subjects of BCI competition III, data set IVA

sensorimotor cortex. Thus, FBCSSP is a convenient method not only for acquiring higher classification rates, but also for extracting the physiological information successfully.

4   Conclusion

In a motor imagery classification problem, applying a band pass filter which is suitable with the frequency band of the subjects sensoriomotor cortex helps finding out better spatial filters that focus on the related area on the motor cortex better. However, the frequency band of the filter is subjective. Thus, searching for a method that automatically sets the required band pass filter is an important issue. Proposed FBCSSP method is a spatio spectral filtering algorithm which optimizes spatial and spectral filters specific to any subject.

FBCSSP method is formed with a filter bank and two consecutive CSP layers in which the first CSP layer plays role on localizing spatial filters specific to a given frequency band while second one weights frequency bands and designs the spectral filter by linearly combining the output of the first layer. The proposed algorithm uses CSP, which is a state of art method in motor imagery classification. Proposed method was inspected in terms of classification performance and physiological plausibility of the obtained spectral and spatial filters. For evaluation, we used a publicly available data set which is very popular in motor imagery classification studies. Classification performance table shows that FBC-SSP algorithm is a successful method. Furthermore, we confirmed the physiological plausibility of the filter by inspecting filters spectral and spatial responses. Reported results show that the FBCSSP method is a successful spatio spectral method for motor imagery signal classification.

Developed method will be adapted for multiple classes as a future work. Also, proposed method's performance will be inspected by testing with more datasets and more spatio-spectral methods found in the literature.

References

1. Ang, K.K., Chin, Z.Y., Zhang, H., Guan, C.: Filter bank common spatial pattern (FBCSP) in brain-computer interface. In: IEEE International Joint Conference on Neural Networks 2008, IJCNN 2008 (IEEE World Congress on Computational Intelligence), pp. 2390–2397. IEEE (2008)
2. Barachant, A., Bonnet, S., Congedo, M., Jutten, C.: Classification of covariance matrices using a riemannian-based kernel for BCI applications. Neurocomputing **112**, 172–178 (2013)
3. Blankertz, B., Muller, K.R., Curio, G., Vaughan, T.M., Schalk, G., Wolpaw, J.R., Schlogl, A., Neuper, C., Pfurtscheller, G., Hinterberger, T., et al.: The BCI competition 2003: progress and perspectives in detection and discrimination of eeg single trials. IEEE Trans. Biomed. Eng. **51**(6), 1044–1051 (2004)
4. Blankertz, B., Muller, K.R., Krusienski, D.J., Schalk, G., Wolpaw, J.R., Schlogl, A., Pfurtscheller, G., Millan, J.R., Schroder, M., Birbaumer, N.: The BCI competition iii: validating alternative approaches to actual BCI problems. IEEE Trans. Neural Syst. Rehabil. Eng. **14**(2), 153–159 (2006)
5. Blankertz, B., Tomioka, R., Lemm, S., Kawanabe, M., Muller, K.R.: Optimizing spatial filters for robust EEG single-trial analysis. IEEE Sig. Process. Mag. **25**(1), 41–56 (2008)
6. Dornhege, C., Blankertz, B., Curio, G., Muller, K.: Boosting bit rates in noninvasive EEG single-trial classifications by feature combination and multiclass paradigms. IEEE Trans. Biomed. Eng. **51**(6), 993–1002 (2004)
7. Dornhege, G., Blankertz, B., Krauledat, M., Losch, F., Curio, G., Muller, K.R.: Combined optimization of spatial and temporal filters for improving brain-computer interfacing. IEEE Trans. Biomed. Eng. **53**(11), 2274–2281 (2006)
8. Falzon, O., Camilleri, K.P., Muscat, J.: The analytic common spatial patterns method for EEG-based BCI data. J. Neural Eng. **9**(4), 045009 (2012)

9. Higashi, H., Tanaka, T.: Simultaneous design of FIR filter banks and spatial patterns for EEG signal classification. IEEE Trans. Biomed. Eng. **60**(4), 1100–1110 (2013)
10. LaFleur, K., Cassady, K., Doud, A., Shades, K., Rogin, E., He, B.: Quadcopter control in three-dimensional space using a noninvasive motor imagery-based brain-computer interface. J. Neural Eng. **10**(4), 046003 (2013)
11. Lemm, S., Blankertz, B., Curio, G., Müller, K.R.: Spatio-spectral filters for improving the classification of single trial EEG. IEEE Trans. Biomed. Eng. **52**(9), 1541–1548 (2005)
12. McFarland, D.J., McCane, L.M., David, S.V., Wolpaw, J.R.: Spatial filter selection for EEG-based communication. Electroencephalogr. Clin. Neurophysiol. **103**(3), 386–394 (1997)
13. Novi, Q., Guan, C., Dat, T.H., Xue, P.: Sub-band common spatial pattern (SBCSP) for brain-computer interface. In: 3rd International IEEE/EMBS Conference on Neural Engineering 2007, CNE 2007, pp. 204–207. IEEE (2007)
14. Pfurtscheller, G., Brunner, C., Schlögl, A., Da Silva, F.L.: Mu rhythm (de) synchronization and EEG single-trial classification of different motor imagery tasks. NeuroImage **31**(1), 153–159 (2006)
15. Pfurtscheller, G., Da Silva, F.L.: Event-related EEG/MEG synchronization and desynchronization: basic principles. Clin. Neurophysiol. **110**(11), 1842–1857 (1999)
16. Ramoser, H., Muller-Gerking, J., Pfurtscheller, G.: Optimal spatial filtering of single trial EEG during imagined hand movement. IEEE Trans. Rehabil. Eng. **8**(4), 441–446 (2000)
17. Tomioka, R., Dornhege, G., Nolte, G., Blankertz, B., Aihara, K., Müller, K.R.: Spectrally weighted common spatial pattern algorithm for single trial EEG classification. Department of Mathematical Engineering, University of Tokyo, Tokyo, Japan, Technical report 40 (2006)
18. Wolpaw, J.R., Birbaumer, N., McFarland, D.J., Pfurtscheller, G., Vaughan, T.M.: Brain-computer interfaces for communication and control. Clin. Neurophysiol. **113**(6), 767–791 (2002)
19. Yuksel, A., Olmez, T.: A neural network-based optimal spatial filter design method for motor imagery classification. PloS one **10**(5), e0125039 (2015)
20. Zhou, S.M., Gan, J.Q., Sepulveda, F.: Classifying mental tasks based on features of higher-order statistics from EEG signals in brain-computer interface. Inf. Sci. **178**(6), 1629–1640 (2008)

Adaptive Segmentation Optimization for Sleep Spindle Detector

Elizaveta Saifutdinova[1,3(✉)], Martin Macaš[2], Václav Gerla[2],
and Lenka Lhotská[2]

[1] Faculty of Electrical Engineering, Department of Cybernetics,
Czech Technical University in Prague, Prague, Czech Republic
saifueli@fel.cvut.cz
[2] Czech Institute of Informatics, Robotics and Cybernetics,
Czech Technical University in Prague, Prague, Czech Republic
[3] National Institute of Mental Health, Prague, Czech Republic

Abstract. Segmentation is a crucial part of the signal processing as it
has a significant influence on further analysis quality. Adaptive segmentation based on sliding windows is relatively simple, works quite good
and can work online. It has however many tunable parameters whose
proper values depend on the task and signal type. The paper proposes a
method of defining optimal parameters for detection of sleep spindles in
electroencephalogram. Segmentation algorithm based on Varri method
was utilized. Fitness function was proposed for estimation of agreement
between the segmentation result and borders of the target classification.
Particle swarm optimization was used to find optimal parameters. On the
data of 11 insomniac subjects the method reached 28 % improvement in
comparison to the baseline method using default parameters.

Keywords: Sleep EEG · Adaptive segmentation · Optimization · Sleep
spindles · Particle swarm optimization

1 Introduction

Signal segmentation is an important step of many signal processing applications.
The task of signal segmentation consists of splitting a non-stationary signal into
quasi-stationary epochs. This is particularly important and frequently used for
long-term electroencephalogram (EEG) analysis. One of the typical problems in
EEG analysis is detection of important EEG patterns like epileptic seizures or
sleep spindles. Typically, pattern classification is applied on features extracted
from short segments [1,13,18,21]. Therefore, the segmentation directly affects
the quality of further signal analysis. Segment borders should correspond to the
borders of the EEG patterns detected by expert as much as possible, otherwise
it would be difficult for the classifier to detect that pattern.

There are two types of segmentation: constant and adaptive. Constant segmentation divides the signal into pieces with the same length, whereas adaptive

© Springer International Publishing Switzerland 2016
M.E. Renda et al. (Eds.): ITBAM 2016, LNCS 9832, pp. 85–96, 2016.
DOI: 10.1007/978-3-319-43949-5_6

one adjusts the segment size to the signal change points. The paper focuses on methods of adaptive segmentation. Adaptive segmentation approaches divide the signal into segments within which the statistics do not change too much. There is a huge family of algorithms based on calculating metrics using sliding window technique. The most popular algorithm is named after its author Alpo Värri [19] and it is based on computing frequency and amplitude measures in two successive sliding windows. A more complex algorithm uses a fractal dimension of the signal in a sliding window as a feature for segmentation [3,5]. Other methods utilize non-linear energy operator [2,10]. These methods have linear complexity, can work online and are very suitable for long-term signal analysis. Other approaches are based on the signal prediction. If a mismatch between the prediction and the original signal is higher than a defined threshold then it suggests a potential segment boundary. Adaptive approaches have many advantages to the constant ones, however the constant segmentation is usually used in patterns detection because of its easy implementation [9].

The sleep spindle can be characterized as sinusoidal wave with 9–16 Hz frequency and 0.5–3 s duration [7]. Sleep spindles play a significant role in modern neuroscience. Along with K-complex they hallmark the second non-rapid eye movement (NREM) sleep stage [7], but can occur in 3rd or 4th NREM sleep stage (according to Rechtschaffen-Kales staging methodology [17]). Moreover, they are connected with cognitive capabilities such a memory consolidation [7,20]. Also the density of sleep spindles (number of events per minute) is an important index in studies of brain and psychological disorders like schizophrenia [6,20], epilepsy [16], and autism [15].

In the most real bio-medical signals it is not an easy task to define clearly the borders between patterns, sometimes they are just unknown. Therefore the majority of studies uses the simulated signals such as [4,14] for segmentation results evaluation. There are two possibilities to measure goodness of segmentation: evaluation based on segments [2] and evaluation based on segment boundaries positions [8]. The sleep spindles are defined by their boundaries and the latter type of evaluation is more natural. It has an intuitively clear way of calculation of goodness of fit using statistical measures of the performance of a binary classification test such as F1 score and Matthews correlation coefficient.

This paper is devoted to developing a method of optimization of adaptive segmentation in sleep spindle detection task. For that purpose a suitable fitness function was proposed, it takes into consideration requirements on the segmentation results. The method is tested on the real EEG data of insomniac subjects. An analysis of parameters optimized on subsets of those data is provided.

2   Materials and Methods

2.1   Dataset

To evaluate the method sufficiently, whole night polysomnography data of 11 patients with main diagnosis of non-organic insomnia (5 women and 6 men) were measured. The age of subjects is between 29 and 53, average age is 41.5.

Data was recorded in National Institute of Mental Health, Prague. Sleep stages were evaluated by trained clinician. EEG was recorded with C3 and C4 against Cz electrodes, the sampling rate is 250 Hz. For future analysis the mean value of channel C3-Cz and C4-Cz was utilized.

In this way, a dataset was obtained, which is large enough for the our particular task of evaluation of segmentation results. Concretely, total analyzed time was 26 h and 4 min, which included 3167 sleep spindles corresponding to 6334 segment borders. Spindles were found manually for these segments by expert. It should be noted, that only parts of the signal with NREM 2 sleep stage were used in the paper.

2.2 Method Overview

The method scheme is introduced on Fig. 1. In the first step the raw signal was filtered using band-pass filter in range 9–16 Hz. Further, the entire dataset was divided into training and validation subsets. Optimization was performed on the training set using particle swarm optimization (PSO). In order to compute the fitness value, the candidate parameters are used in the segmentation described in the Sect. 2.3 and the results are evaluated based on agreement with manually labeled spindles boundaries. The fitness function measures how close the obtained segmentation and the expert signal labels are (see Sect. 2.5).

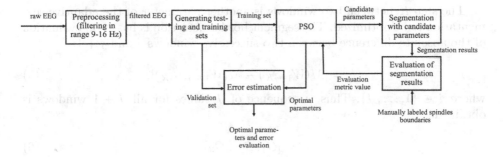

Fig. 1. Method overview.

2.3 Segmentation

Prior to the segmentation itself, appropriate preprocessing was applied to the signal. It was filtered in frequency range associated with sleep spindles (9–16 Hz). Further, segment boundaries are detected by detecting changes in the signal using a sliding window method [14]. In this step two successive windows are sliding along the signal and for each window a metric is computed. The Varri's method [14,19] is used to compute such metric.

Let $[x_1, x_2, x_3, \ldots, x_{N-1}, x_N]$ be the signal of length N and let $W_i = [x_i, x_i + 1, x_i + 2, \ldots, x_{i+L-1}]$ be a window of length L starting at each sample of the signal. The frequency measure for window W_i is defined as

$$F(W_i) = \sum_{j=1}^{L-1} |x_{i+j} - x_{i+j-1}| \tag{1}$$

and the amplitude measure is

$$A(W_i) = \sum_{j=1}^{L} |x_{i+j-1}|. \tag{2}$$

A function M evaluates a window W_i using a combination of the frequency and amplitude measures:

$$M(W_i) = k_1 F(W_i) + k_2 A(W_i), \tag{3}$$

where k_1 and k_2 are coefficients weighting the amplitude and frequency components. Those coefficient are parameters of the segmentation that are optimized in our experiments. Further, one window is evaluated by M every K steps so that one obtains $J + 1$ values

$$[M(W_1), M(W_{1+K}), M(W_{1+2K}), \ldots, M(W_{1+JK})]. \tag{4}$$

The step K between two windows is another parameter of the adaptive segmentation to be optimized. The segment border detection is based on evaluation of the absolute difference between two successive windows j and $j + 1$:

$$G_j = |M(W_{1+jK}) - M(W_{1+(j-1)K})|, \tag{5}$$

where $j \in \{1, \ldots, J\}$. Thus, a sequence of J values for all $J + 1$ windows is obtained:

$$G = [G_1, G_2, ..G_J], \tag{6}$$

which reflects the change of statistical properties of the signal. G is further normalized by division of each G_j by $\max_j G_j$. The border detection is performed by thresholding the G, detecting local maxima, and simultaneously satisfying two other constraints. The distance between the peaks must be higher th the window size L and the amplitude of the peak must be higher than standard deviation of the thresholded signal. Adaptive threshold is used, which is obtained as a moving average value of the G sequence multiplied by threshold coefficient c. The size of the moving average window and the threshold coefficient are also parameters to be optimized. In Fig. 2 one can see an example of the raw and the filtered signals, G sequence and corresponding segmentation.

The window length parameter L is important. It should be large enough to detect difference in the two windows at all, but it should not be too large, which could avoid to detect some borders by capturing lot of real segments

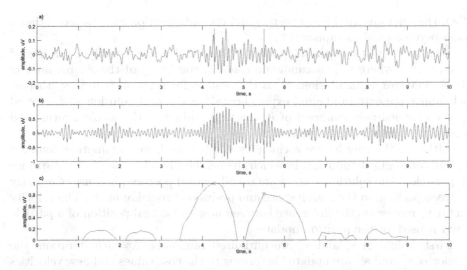

Fig. 2. Segmentation process. From top to bottom: (a) original raw EEG data; (b) signal filtered in range 9–16 Hz; (c) G function based on Varri metric of filtered signal with labeled as red circles maximum of the function. Red lines show the segmentation obtained using G function. (Color figure online)

in one window. The proper window length can be chosen using energy of the G sequence. Ideally, G function with the proper window size should be highly above zero close to segment boundaries and almost zero elsewhere. This property can be evaluated using energy value. For an improper window length, G function has more energy compared to a proper window length. Thus, for an analyzing window with length L, the energy of the G sequence can be calculated as its mean squared value:

$$E = \frac{\sum_{j=1}^{J} G_j{}^2}{J}, \qquad (7)$$

where J is the length of the G sequence. A proper window length should correspond to the minimum of the energy curve. In the paper energy of the corresponding G function was calculated for optimal segmentation and segmentation with default parameters.

2.4 Optimization Using Particle Swarms

The six parameters of the adaptive segmentation defined above were optimized by the particle swarm optimization (PSO) method. It is one of optimization methods developed for finding a global optimum of some nonlinear function [11]. It has been inspired by social behavior of birds and fish. The method applies the approach of problem solving in groups. Each solution consists of a set of parameters and represents a position in multidimensional space. In continuous PSO, the solutions move in the search space in a swarm-like group. In the binary

PSO, the solutions are binary vectors and the optimization process has rather a socio-psychological metaphor [12].

The main entity of the algorithm is a particle. Each particle consists of $\mathbf{s}_i, \mathbf{v}_i, \mathbf{s}_i^P, \mathbf{s}_i^G$, where $\mathbf{s}_i$ is a candidate solution consisting of the six parameters of the adaptive segmentation, $\mathbf{s}_i^P$ is the best so far solution found by ith particle and represents individual experience. $\mathbf{s}_i^G$ is the best solution so far found by a predefined neighborhood of ith particle (subset of the whole swarm) and represents the social knowledge.

Although the ring lattice sociometry is often used, we use another common setting called gBest topology, in which the neighborhood is the whole swarm for all particles. The solution vectors are usually called position, because of analogy between position in the search space and position of social animals. The velocity vector $\mathbf{v}_i$ represents the difference between new and actual position of a particle i and is used for the position update.

First, values of $\mathbf{s}_i$ and $\mathbf{v}_i$ are initialized randomly. At each iteration, the memories $\mathbf{s}_i^P$ and $\mathbf{s}_i^G$ are updated according to the cost values and new velocities are computed. Finally, the position update is performed. In the original version of the PSO, the velocity update is accomplished by the following equation:

$$\mathbf{v}_i(t+1) = \omega \mathbf{v}_i(t) + \varphi_1 \mathbf{R}_1(\mathbf{s}_i^P(t) - \mathbf{s}_i(t)) + \varphi_2 \mathbf{R}_2(\mathbf{s}_i^G(t) - \mathbf{s}_i(t)), \qquad (8)$$

where the symbols $\mathbf{R}_1$ and $\mathbf{R}_2$ represent the diagonal matrices with random diagonal elements drawn from a uniform distribution between 0 and 1. The parameters φ_1 and φ_2 are scalar constants that weight influence of particles' own experience and the social knowledge. The parameter ω is called inertia weight and its behavior determines the character of the search. Further, new candidate solutions (new positions) are updated. For continuous optimization, the position update is simple:

$$\mathbf{s}_i(t+1) = \mathbf{s}_i(t) + \mathbf{v}_i(t+1). \qquad (9)$$

At each step, the PSO algorithm modifies the distance that each particle moves on each dimension. Changes in the velocity are stochastic, and it can cause an undesirable expansion of particles trajectory into wider and wider cycles [12]. One solution is to implement boundaries for the velocity. If any component of $\mathbf{v}_i$, v_i^a is lower than $-v_{max}^a$ or greater than $+v_{max}^a$, its value is replaced by $-v_{max}^a$ or $+v_{max}^a$, respectively. Note that there is different maximum velocities for different components of the velocity vectors. The maximum velocity was set as the range of the search space in the particular component.

Finally, a constraint handling technique was used to keep the parameter values within their feasible range. If any component of the position vector gets out of the range, it is re-initiated on the border of the range and the corresponding component of the velocity is inverted. Thus, the particles are repelled from the constraint barrier.

Within the experiments, we used the following setting for PSO parameters. The weight parameters $\varphi_1 = \varphi_2 = 2.1$, the inertia was linearly decreasing from

0.6 to 0.3, swarm size was 25 and the number of iterations was 100. Those parameters were not tuned as they correspond to typical and most common setting in literature.

2.5   Fitness Function

In the paper the real EEG data with manually labeled sleep spindles were used. It could be taken as the target segmentation, but the segment boundaries are not always so unambiguous. For this reason instead of real borders as a gold standard segmentation we used set of ranges where spindle border could occur like it was in [8]. Each range is neighborhood of the real spindle border of size 0.125 s, the size of the neighborhood was chosen empirically. The minimal length of spindle is 0.5 s, so the ranges do not overlap.

Since the target segmentation is determined, the true positives (TP) could be defined as number of ranges where there is at least one of found segmentation borders and false negatives (FN) as number of ranges where there is no segmentation borders. Sum of TP and FN equals the number of ranges in the target segmentation.

Moreover, spindles are not distributed equally in the whole signal. So, it can happen that segmentation algorithm finds additional borders, but it could be connected with signal change because of other processes. Those false positive however do not matter if non-spindle segments are long enough. Assuming that segment is a signal part between two successive borders and minimal length of the segment equals the minimal spindle length (0.5 s), penalty value (PV) has been introduced and it equals the number of segments shorter than the minimal length. TP, FN and PV examples are presented on Fig. 3.

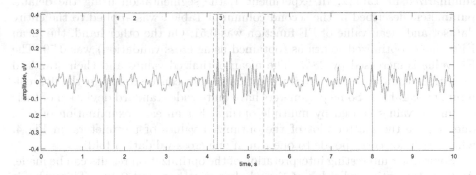

Fig. 3. Examples of TP, FN and PV. The red rectangles represent the ranges where the real spindle borders occur, the blue lines are the result of segmentation. Line number 3 represents TP and absence of any found borders in the second range points on FN. Distance between lines 1 and 2 is less than 0.5 s which increases PV by 1 point. And line number 4 stands further than 0.5 s from other borders, it does not change PV value. (Color figure online)

Aggregating all those requirements on the result of segmentation, fitness function (FF) is defined as:

$$FF = \frac{2TP}{2TP + FN + PV}. \tag{10}$$

The FF aggregates TP, FN and PV by analogy with f-measure in statistical analysis of binary classification and measures an agreement with expert segmentation on spindles and non-spindles segments. It varies from 0 when there is empty TP set to 1 where there are no FN and PV.

3   Empirical Results

3.1   Experimental Settings

For the experimental results data of 11 subjects was used. The data was divided into 5 folds, each fold corresponds to the data of 2 or 3 subjects. Optimization of parameters described in Table 1 was tested by leave-one-out cross validation scheme. This means that the optimization procedure was repeated for each fold as a validation set and all the remaining folds as a training data. The optimal parameters and their performance values were obtained 5 times, i.e. one for each fold.

3.2   Results

The default and optimized parameter values and their feasible ranges are summarized in Table 1. The FF values averaged over the cross-validation folds are summarized in Table 2. In experiment 1, the segmentation using the default parameters described in the second column of Table 1 was applied to the entire dataset and mean value of FF function was 0.51. On the other hand, the mean FF value of optimized solutions (obtained using cross validation) was 0.79. The FF value is increased by 28 %. The mean optimized values and their standard deviations computed over cross-validation results can be found in the 4th column of the Table 1. Some parameters have quite wide standard deviation of their optimized values caused by multiple optima. For an easy examination of this, one can see the scatter plot of the optimized values of parameters on Fig. 4, where each dot corresponds to one run in one cross-validation fold.

Some other interesting interpretation of the optimization results can be made. At first, the optimized value of the window size is around 0.48 s. This value is very close to a minimal expected length of the spindle.

An obvious question is, if it would be sufficient to optimize only the window size and let all the other parameters default. The comparison is made in Table 2. In experiment 3, adaptive segmentation with the window size $L = 0.48$ s, but all the other parameters default, was used. The average fitness value equals to 0.72, which is outperformed by the fully optimized adaptive segmentation and thus it makes sense to optimize all the parameters.

Table 1. Summary of parameters of the adaptive segmentation, their default values, range and optimized value

Parameter	Default	Range	Optimized
Window length (seconds)	1	$\langle 0.25, 3 \rangle$	0.48 ± 0.01
Step K (signal samples)	24	$\langle 5, 56 \rangle$	5.1 ± 0.1
Threshold window (G sequence samples)	5	$\langle 2, 20 \rangle$	5.16 ± 3.67
Threshold coefficient	1	$\langle 0, 1 \rangle$	0.52 ± 0.56
k1	1	$\langle -100, 100 \rangle$	-13.52 ± 17.27
k2	7	$\langle -100, 100 \rangle$	-21.8 ± 84.96

Table 2. Summary of experiments and average fitness function values

Experiment	Approach	FF value
1	Adaptive, default	0.51
2	Adaptive optimized	0.79
3	Adaptive, L=0.48, other parameters default	0.72
4	Constant, L=1	0.41
5	Constant, L=0.48	0.67

A similar setting, but for constant segmentation, was performed in experiments 4 and 5, where a constant segmentation with $L = 1$ obviously leads to the worst result ($FF = 0.41$) and increases significantly to $FF = 0.67$ if the optimized window size $L = 0.48$ s is used. Nevertheless, the adaptive segmentation with all parameters optimized simultaneously is still much better choice. This conclusion correspond to the difference of energy of G sequence defined in Eq. 7. While in experiment 1, the energy is 0.056, in experiment 2 it is 0.037, which confirms that the window size 0.48 s is better choice for the segmentation.

Since adaptive threshold was used in the presented method, there are two parameters representing it. The threshold window size is a parameter of moving average filter of the G function and the threshold coefficient which scales the threshold relatively to standard deviation of the G sequence. In Fig. 4, one can see a strong interaction between threshold coefficient and the threshold window causing two different local optima. First one is that the coefficient is very close to 0 and the window size is wider than 5 s. It gives no threshold and local peaks are looking in the function G. The second optimum is very close to 1 but the threshold window is quite narrow (about 2.17 ± 0.15 s). In this case, the adaptive threshold looks like the G function and only the highest peaks are above the threshold.

Concerning the parameters k_1 and k_2, there is a tendency that absolute value of k_2 should be greater than absolute value of k_1 in average on 62.17 ± 24. This can be partly caused by the fact that the band-pass filtering was used for

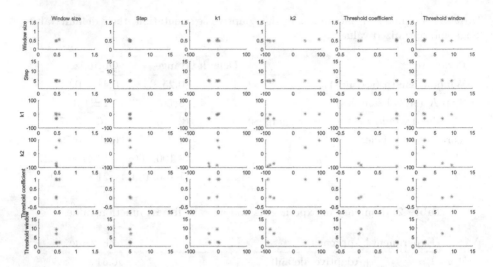

Fig. 4. Optimal parameters obtained in cross validation. Each figure represented distribution of one parameter against the other.

preprocessing, which reduces the relative importance of the frequency measure term.

The optimized value of the window step K was about 5 samples, which is its minimal admissible value. That means that the point of the signal is tested for being a segment border every 5 steps.

It should be noticed that the other important segmentation is an execution time and space complexity, which are directly influenced by the window size and the step. The smaller values of these parameters lead to the higher execution duration and the biggest memory allocation. This aspect is however not considered in our experiments and the optimization does not take the temporal and space complexity into account. Since the swarm size was set to 20 and the number of iterations was 150, FF evaluation (segmentation and evaluation of the results) is performed 20*150 times in one run of PSO, which can make the method time consuming and it could be worth to use some additional penalization of candidate solutions with high complexity of FF evaluation.

4   Conclusion and Discussion

The method to optimize adaptive segmentation for the sleep spindle detection task was presented in the paper. The method is based on the PSO optimization algorithm and maximizes agreement measure between the results of segmentation and manually labeled spindles in the real EEG data. By the cross-validation technique it was shown that using optimized parameters give 28 % higher agreement value than default ones. Obtained optimal parameters were analyzed and the optimal size of a sliding window was found and equals to 0.48. An energy of G sequence was compared for the segmentation with different window size and a reduction of the energy for optimized value against the default one was observed. This points out the proper window size for the spindle segmentation task. It turns out that using that window might not require the threshold and that it leads to increase of the execution time. In the Varri metric it was proved that impact of amplitude measure should be greater than impact of frequency measure.

Future work will be dedicated to the research of the parameters: stability investigation, looking for other minimums of the function and connected parameter clusters, patterns in data. The question about optimal parameters which do not lead to the time and space consuming segmentation process is still open and extended parameters analysis could help with that issue.

In future, the impact of the method for classification of the sleep spindles and its impact on the classification performance will be focused. In fact, it is possible to optimize segmentation using the misclassification rate. Besides the sleep spindles there are another interesting EEG patterns such as K-complexes and proposed method could be applied for the automatic identification of others EEG patterns.

Acknowledgments. Research of E. Saifutdinova was supported by the project No. SGS16/231/ OHK3/3T/13 of the Czech Technical University in Prague. Research of Martin Macas was supported by the project No. GP13-21696P "Feature selection for temporal context aware models of multivariate time series" of the Grant Agency of the Czech Republic (GACR). This publication was supported by the project "National Institute of Mental Health (NIMH-CZ)", grant number CZ.1.05/2.1.00/03.0078 and the European Regional Development Fund.

References

1. Acr, N., Gzeli, C.: Automatic recognition of sleep spindles in EEG by using artificial neural networks. Expert Syst. Appl. **27**(3), 451–458 (2004)
2. Agarwal, R., Gotman, J.: Adaptive segmentation of electroencephalographic data using a nonlinear energy operator. In: Proceedings of IEEE International Symposium on Circuits and Systems (ISCAS 1999), Orlando, FL, vol. 4, pp. 199–202 (1999)
3. Anisheh, S.M., Hassanpour, H.: Adaptive segmentation with optimal window length scheme using fractal dimension and wavelet transform. IJE Trans. B: Appl. **22**(3), 257–268 (2009)
4. Appel, U., Brandt, A.: A comparative study of three sequential time series segmentation algorithms. Sig. Process. **6**(1), 45–60 (1984)
5. Esteller, R.: A comparison of waveform fractal dimension algorithms. Circuits Syst. I: Fundam. Theory Appl. **48**(2), 177–183 (2001)
6. Ferrarelli, F., et al.: Reduced sleep spindle activity in schizophrenia patients. Am. J. Psychiatry **164**(3), 483–492 (2007)
7. Gennaro, L.D., Ferrara, M.: Sleep spindles: an overview. Sleep Med. Rev. **7**(5), 423–440 (2003)
8. Gensler, A., Sick, B.: Novel criteria to measure performance of time series segmentation techniques. In: LWA (2014)

9. Gunes, S., Dursun, M., Polat, K., Yosunkaya, S.: Sleep spindles recognition system based on time and frequency domain features. Expert Syst. Appl. **38**(3), 2455–2461 (2011)
10. Kaiser, J.F.: On a simple algorithm to calculate the 'energy', of a signal. In: Proceedings of IEEE International Conference on Acoustics, Speech, and Signal Processing, Albuquerque, NM, vol. 1, pp. 381–384 (1990)
11. Kennedy, J., Eberhart, R.: Particle swarm optimization. IEEE International Conference on Neural Networks 1995. Proceedings, vol. 4, pp. 1942–1948 (1995)
12. Kennedy, J., Eberhart, R.C.: Swarm Intelligence. Morgan Kaufmann Publishers Inc., San Francisco (2001)
13. Koley, B., Dey, D.: An ensemble system for automatic sleep stage classification using single channel eeg signal. Comput. Biol. Med. **42**(12), 1186–1195 (2012)
14. Krajca, V., Petranek, S., Patakova, I., Varri, A.: Automatic identificaton of significant graphoelements in multichannel eeg recordings by adaptive segmentation and fuzzy clustering. Int. J. Biomed. Comput. **28**, 71–89 (1991)
15. Limoges, E., Mottron, L., Bolduc, C., Berthiaume, C., Godbout, R.: Atypical sleep architecture and the autism phenotype. Brain: J. Neurol. **128**(5), 1049–1061 (2005)
16. Myatchin, I., Lagae, L.: Sleep spindle abnormalities in children with generalized spike-wave discharges. Pediatr. Neurol. **36**(2), 106–111 (2007)
17. Rechtschaffen, A., Kales, A.: A manual of standardized terminology, techniques and scoring system for sleep stages of human subjects. Clin. Neurophysiol. **26**(6), 644 (1968)
18. Sinha, R.K.: Artificial neural network and wavelet based automated detection of sleep spindles, rem sleep and wake states. J. Med. Syst. **32**(4), 291–299 (2008)
19. Varri, A.: Digital Processing of the EEG in Epilepsy. Ph.D. Thesis, Tampere University of Technology, Tampere, Finland (1988)
20. Wamsley, E.J., et al.: Reduced sleep spindles and spindle coherence in schizophrenia: mechanisms of impaired memory consolidation? Biol. Psychiatry **71**(2), 154–161 (2012)
21. Zoubek, L., Charbonnier, S., Lesecq, S., Buguet, A., Chapotot, F.: Feature selection for sleep/wake stages classification using data driven methods. Biomed. Sig. Process. Control **2**(3), 171–179 (2007)

Probabilistic Model of Neuronal Background Activity in Deep Brain Stimulation Trajectories

Eduard Bakstein[1(✉)], Tomas Sieger[1,2], Daniel Novak[1], and Robert Jech[2]

[1] Department of Cybernetics, Faculty of Electrical Engineering,
Czech Technical University in Prague, Prague, Czech Republic
eduard.bakstein@fel.cvut.cz
[2] Department of Neurology and Center of Clinical Neuroscience,
First Faculty of Medicine and General University Hospital,
Charles University in Prague, 128 21 Prague, Czech Republic

Abstract. We present a probabilistic model for classification of micro-EEG signals, recorded during deep brain stimulation surgery for Parkinson's disease. The model uses parametric representation of neuronal background activity, estimated using normalized root-mean-square of the signal. Contrary to existing solutions using Bayes classifiers or Hidden Markov Models, our model uses smooth state-transitions represented by sigmoid functions, which ensures flexible model structure in combination with general optimizers for parameter estimation and model fitting. The presented model can easily be extended with additional parameters and constraints and is intended for fitting of a 3D anatomical model to micro-EEG data in further perspective. In an evaluation on 260 trajectories from 61 patients, the model showed classification accuracy 90.0 %, which was comparable to existing solutions. The evaluation proved the model successful in target identification and we conclude that its use for more complex tasks in the area of DBS planning and modeling is feasible.

Keywords: Deep brain stimulation · Microelectrode recordings · Probabilistic model

1 Introduction

The Deep Brain Stimulation (DBS), which consists of permanent electrical stimulation of the basal ganglia, has been used for treatment of Parkinson's disease (PD) and other movement disorders since the pioneering work by Benabid et al. in the early 1990s [4]. Since then, it has become a standard therapy for drug-resistant late-stage Parkinson's disease and is applied in hundreds of centers worldwide. In order to achieve a good clinical outcome, accurate positioning of the stimulation electrode is necessary. As the target structures are small — the most common target for PD, the subthalamic nucleus (STN) measures less than 10 mm along its largest dimension — and precision of imaging methods available prior to operation is relatively low (~ 1 mm voxel size in pre-operative magnetic

© Springer International Publishing Switzerland 2016
M.E. Renda et al. (Eds.): ITBAM 2016, LNCS 9832, pp. 97–111, 2016.
DOI: 10.1007/978-3-319-43949-5_7

resonance imaging scans), precise placement based solely on pre-operative imaging is hard.

To obtain a more accurate location information, a method called microrecording is employed in a vast majority of DBS centers [1]. In microrecording, a set of microelectrodes (tip diameter around 5 μm) is shifted through the brain and microelectrode EEG (also μEEG or MER) is recorded. The recorded signals are evaluated concurrently by a trained neurologist, who then identifies optimal position for the stimulation contacts. The evaluation is typically based on visual and auditory inspection of the signals, the main markers being neuronal firing pattern and especially amplitude of the neuronal background, which are higher in areas with higher neuron density — such as the STN. The accumulation of neurons in the STN is very high compared to the neighboring structure, which projects into the recorded signals as a sudden increase in the neuronal background activity as the electrode approaches the STN boundary, as well as appearance of rapidly spiking neurons once the electrode entered the nucleus. The former can be estimated by the root mean square (RMS) of the original signal [9,14], some authors also suggested signal with removed spikes or RMS of a band-pass filtered signal [10,11].

For a long time, efforts have been made to use machine learning models in place of the manual evaluation. This paper presents a probabilistic model of neuronal background activity along a microrecording trajectory, characterized by a normalized root-mean-square measure (NRMS). The suggested model is a logical extension of already existing models, which are summarized in the next section.

1.1   Existing Models

Early models used the neuronal background level, estimated using the normalized root-mean-square of the signal as an input to Bayesian classifier [9] or discrete hidden Markov model (HMM) [14]. These models included also the expected distance to target as an input, which utilizes the fact that the pre-surgical planning places the target (i.e. "depth 0") to a specific part of the STN. These models also used manual quantization or thresholding of the input parameters in order to achieve reasonably-sized discrete parametric space, that can be estimated from commonly-sized training datasets.

Extension to semi-markov models, including state duration (i.e. the length of nuclei pass) with continuous probability density function has been done by Taghva et al. [13], but has been evaluated only on simulated data. Other researchers investigated features such as high-frequency component of the neuronal background [10] or multiple features including power spectral density, firing rate and noise level coupled with a rule-based classifier composed of cascaded thresholds [5]. Support vector machine classifier on multiple signal features (RMS, nonlinear energy, curve length, zero crossings, standard deviation and number of peaks) has been also implemented by Guillen et al. [7] with almost 100 % accuracy.[1]

The authors of [12] investigated the impact of recording length and density on performance of an HMM and concluded that precision of a previously published HMM model [14] was approximately half of the between-position distance.

1.2 Proposed Model

In this paper, we present a model based on the neuronal background level, which can be used as a basis for fitting anatomical 3D model directly to the recorded μEEG activity along parallel trajectories. The presented variant is a one-dimensional proof of concept, intended to verify the idea and compare its properties to existing well-performing models.

Similarly to the hidden semi-markov models used in [13], our model uses parametric representation of input feature space – the NRMS values computed according to [9] but without quantization. Contrary to HMM, our model uses smooth state to state transitions, motivated by properties of electrical field of the STN, observed on the training data.

A derived model, based on the proposed approach, can be used to introduce other requirements such as the expected length of STN pass for given trajectory, based on a-priori information from surgical planning. Owing to the smooth state transitions, the model has also a smooth likelihood function (and gradient) and can be fitted using general purpose optimization algorithms. Thanks to this property, the structure of the model is very flexible and can be easily modified and extended. Moreover, the model theoretically allows classification with accuracy beyond the resolution of the measured data. However, this may not be the case practically due to noise in the μEEG signal and other measurement inaccuracies.

2 Methods

The probabilistic model, presented in this paper, is based on the assumption of different distribution of neuronal background before, within and beyond the STN. Each of these distributions is represented parametrically and transitions between the consecutive distributions are modeled by the logistic sigmoid function (see Sect. 2.2 below). In this section, we give overview of the proposed model, as well as of the data collection and pre-processing.

2.1 Data Collection, Annotation and Pre-processing

The experimental dataset was collected during the standard surgical procedure of DBS implantation using a set of one to five tungsten microelectrodes, spaced

---

[1] The dataset in [7] consisted of 52 signals from four patients only and it is not clear whether the validation sample was completely independent in terms of similarity of neighbor segments — see e.g. [8] for description of a similar problem.

2 mm apart in a cross; the so-called Ben-gun configuration [6]. The microelectrode signals were recorded at each 5 mm along the trajectory using the Leadpoint recording system (Medtronic, MN), sampled at 24 kHz, band-pass filtered in the range 500–5000 Hz and stored for offline processing. Annotation of nucleus at each position was done manually by an expert neurologist [R.J.], based on visual and auditory inspection of the recorded signal.

To reduce the effect of motion-induced artifacts, we divided each signal into 1/3 s windows and selected the longest stationary component using the method presented in [3], which is an extension of method previously presented in [2]. Parameters of the method (detection threshold and window length) were selected in order to achieve best accuracy on a training database. This method was chosen in order to obtain at least some segment of each signal, even though it may contain electromagnetic and other interference, which would be marked as signal artifact by the stricter spectral method, presented in [3].

2.2 Electric Field of the STN

To obtain estimate of the neuronal background activity level, we calculated the root-mean-square (RMS) of the stationary portion of the signal. In accordance with [9], we computed the normalized RMS of the signal (NRMS) by dividing feature values of the whole trajectory by mean RMS values of the first 5 positions (which are assumed non-STN in a majority of recordings). Additionally, we normalized the 90th percentile of each NRMS trajectory to 3 in order to limit NRMS variability in the STN.

Observations of NRMS values before, within and after the STN confirmed different distribution in each part. After comparing likelihood of normal and log-normal distribution, we chose to model the NRMS values in each part by the best-fitting log-normal distribution.

Further explorative analysis was aimed at the shape of NRMS transition. Figure 1 presents NRMS training data, aligned around STN entry and exit, mean value for each distance to the transition and the sigmoid logistic function we chose to model the transition as a result.

2.3 Parametric Model of STN Background Activity

Model Structure. The proposed model of background activity along the DBS trajectory consists of probability density of the NRMS measure in the three different regions. These can be seen as continuous emission probabilities in three hidden states of an HMM. Contrary to an HMM, the proposed model uses no discrete state transitions that could be represented by a transition matrix, but uses smooth state transitions, represented by sigmoid (or logistic) functions. Due to that, standard evaluation methods used for HMM, such as the Viterbi algorithm, can not be used and are replaced by general constrained optimization.

The general idea of the proposed model is based on the following reasoning: one of the most obvious features, distinguishing DBS target structure in

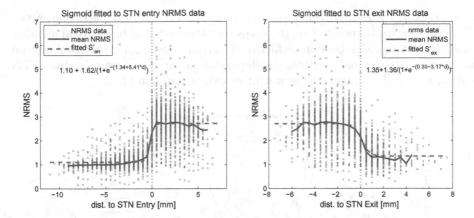

Fig. 1. NRMS values around STN entry and exit points (depth 0 on the x axis) from a set of training trajectories. The blue line represents mean NRMS value for each distance, the red dashed line shows fitted sigmoid functions S'_{en} and S'_{ex}, used to model STN entry and exit transitions, with parameters corresponding to the inlaid formula. (Color figure online)

the μEEG — in particular the STN — is signal power, represented here by signal NRMS. Based on our observations on training trajectories (see Sect. 2.2), as well as previous works (e.g. [10,11]), we assume different probability distribution of NRMS values in the areas before, within and beyond the STN and use the log-normal distribution as a model for the NRMS values in each area. Parameters of the log-normal model are estimated from labeled training data during the training phase.

In common settings, the μEEG signals are recorded at discrete depth steps (in our case every 0.5 mm). The task is therefore to classify signals, recorded at each position, to a correct class (i.e. identify the STN). We assume that the electrode can pass through the STN at most once and the trajectory can thus be divided into three consistent segments by two boundary points: STN entry and STN exit. In the evaluation phase we find optimal STN entry and exit points by maximizing the joint likelihood of the observed NRMS values along the trajectory with respect to the previously identified probability distributions. Simply put, the values before the assumed STN entry should be close to the expected value of the distribution before the STN, the values within the assumed STN should be close to the expected value of the distribution within STN and accordingly for the area beyond STN.

In order to increase theoretical precision of the model, as well as to improve its algebraic properties[2], we add smooth state transitions, modeled using logistic sigmoid functions. This approach also seems to be well in alignment with the observed statistical properties of NRMS values around STN boundary points — as can be

---

[2] Smooth state transitions using logistic sigmoid functions lead to smooth gradient and the resulting model is therefore easier to optimize.

seen in Fig. 1. The result of this addition is that rather than belonging to one particular state, each data point along the trajectory is assumed to be a partial member of all three states. Membership coefficients c_{pre}, c_{STN} and c_{post} of this combination are given by the sigmoid functions and depend on distance of given point from STN entry and exit. Illustration of the weighting can be found in Fig. 2.

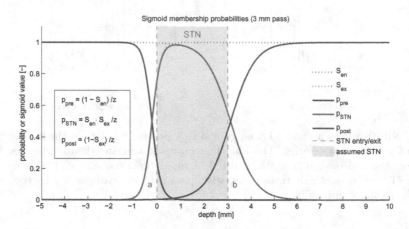

Fig. 2. Illustration of sigmoid transition functions S_{en} and S_{ex} and their application to the joint likelihood function from Eq. 8: each observed data point is assumed to be a partial member of all three hidden states. Probability density functions corresponding to each state are weighted using the membership probabilities $p_{pre}(i) = p(d_i \in pre|a, b, \Theta)$, $p_{STN}(i) = p(d_i \in STN|a, b, \Theta)$ and $p_{post}(i) = p(d_i \in post|a, b, \Theta)$ which are dependent on distance from the hypothetical STN entry and exit points a and b. The $z(i) = z_i$ is normalization coefficient - see Eqs. 10 and 13 for details.

In this paper, we present two variants of the model: (i) the basic *flex1*, based solely on the NRMS measure and (ii) extended model *flex2*, which adds a-priori distribution of expected STN entry and exit depths. The following sections provide formal definition of the model, as well as the training and evaluation procedure.

Training Phase. Supervised model training is performed on NRMS feature values $x_i \in \{x_1, x_2, ..., x_N\}$, extracted from MER data recorded at N recording positions at depths $d_i \in \{d_1, d_2, ..., d_N\}$. Manual expert annotation is provided for each recording position, labeling the signal as either *stn* or *other*. STN entry position i_{en} and exit depth i_{ex} is defined as index of the first and last occurence of *stn* label from the start of the trajectory. Trajectory is then divided into three parts; (i) before the STN with indices $I_{pre} = \langle 1, i_{en} - 1 \rangle$, (ii) within the STN $I_{stn} = \langle i_{en}, i_{ex} \rangle$ and (iii) after the STN $I_{post} = \langle i_{ex} + 1, N \rangle$. Two groups of parameters are fitted during the training phase:

(i) Parameters of the log-normal probability distribution of NRMS feature values before the STN ($\boldsymbol{\theta}_{pre} = \{\hat{\sigma}_{pre}, \hat{\mu}_{pre}\}$), within the STN ($\boldsymbol{\theta}_{stn}$) and after the STN ($\theta_{post}$), where $\hat{\mu}$ and $\hat{\sigma}$ are maximum-likelihood estimates of location and scale parameters of the respective log-normal distribution, computed in standard way according to

$$\hat{\mu}_{pre} = \frac{\sum_{i \in I_{pre}} \ln(x_i)}{n_{pre}} \tag{1}$$

$$\hat{\sigma}_{pre} = \sqrt{\frac{\sum_{i \in I_{pre}} (\ln(x_i) - \hat{\mu}_{pre})^2}{n_{pre}}} \tag{2}$$

where $n_{pre} = |I_{pre}|$, i.e. the number of positions with given label. Parameters for *stn* and *post* labels are computed accordingly on samples from the I_{stn} and I_{post} sets.

(ii) Parameters defining the shape of the sigmoid transition functions at STN entry (β_{en}^0 and β_{en}^1) and exit (β_{ex}^0 and β_{ex}^1). Here, the parameter β^0 represents shift and β^1 steepness of the respective logistic sigmoid function, defined as

$$S'_{en}(d_i) = \alpha_{en}^0 + \alpha_{en}^1 \cdot \left(1 + \exp\left(-(\beta_{en}^0 + \beta_{en}^1(d_i - d_{en}))\right)\right)^{-1} \tag{3}$$

for STN entry and

$$S'_{ex}(d_i) = \alpha_{ex}^0 + \alpha_{ex}^1 \cdot \left(1 + \exp\left(-(\beta_{ex}^0 + \beta_{ex}^1(d_i - d_{ex}))\right)\right)^{-1} \tag{4}$$

for STN exit, where d_{en} is STN entry depth and d_{ex} STN exit depth. The additional parameters α^0 (shift along the y axis) and α^1 (scaling factor) serve to provide sufficient degrees of freedom to achieve appropriate fit. However, these parameters are not part of the model and are not stored as both are replaced by the log-normal probability density functions modeling the NRMS values in the respective area. Note that contrary to shifted and scaled functions S'_{en} and S'_{ex} fitted during the training phase, standard logistic functions S_{en} and S_{ex} from Eqs. 11 and 12 are used during evaluation. Fitting can be done using general purpose optimization function minimizing mean square error on all training data at once, according to:

$$\underset{\alpha_{en}^0, \alpha_{en}^1, \beta_{en}^0, \beta_{en}^1}{\arg\min} \sum_{i \in I_{pre}, I_{stn}} \left(S'_{en}(d_i, \alpha_{en}^0, \alpha_{en}^1, \beta_{en}^0, \beta_{en}^1) - x_i\right)^2 \tag{5}$$

and similarly for S'_{ex}. Only data labeled as *pre* and *stn* are used to fit parameters of S'_{en} and data labeled as *stn* and *post* are used to fit S'_{ex}. Initial parameters are set to $[\alpha_{en}^0, \alpha_{en}^1, \beta_{en}^0, \beta_{en}^1] = [1, 1, 0, 1]$ and $[\alpha_{ex}^0, \alpha_{ex}^1, \beta_{ex}^0, \beta_{ex}^1] = [1, 1, 0, -1]$

The trained model is then completely characterized by parameter vector $\Theta = \{\theta_{pre}, \theta_{stn}, \theta_{post}, \beta_{en}^0, \beta_{en}^1, \beta_{ex}^0, \beta_{ex}^1\}$, encompassing both log-normal emission probabilities and steepness and shift parameters of the sigmoid transition functions. If more trajectories are available for training, both parameter groups are estimated using all training data at once, given that appropriate labels and STN entry and exit depths are applied for each trajectory separately.

Extended Model. The presented model structure uses no prior information about expected STN entry and exit depths. It is possible to modify the model by adding empirical distribution of entry and exit depths, modeled using the normal distribution $p_a = N(\mu_a, \sigma_a)$ and $p_b = N(\mu_b, \sigma_b)$. The parameters can be estimated using the standard maximum likelihood estimates of mean and standard deviation. This will lead to addition of four parameters. We will denote the extended parameter vector Θ', the extended model is then nicknamed *flex2* in the results section.

Model Evaluation. In the evaluation step, the model with parameters Θ is fitted to a trajectory formed by a sequence of feature values x_i measured at corresponding depths d_i. Optimal posterior STN entry and exit points a and b are identified by minimizing the negative log-likelihood function

$$\{a, b\} = \arg\min_{a,b} \sum_{i=1}^{N} -\ln(L(\{x_i, d_i\}|a, b, \Theta)) \tag{6}$$

The joint likelihood for position i at fixed values of STN entry and exit depths a and b and all three possible states (*pre*, *STN* and *post*) is given by:

$$\begin{aligned} L(\{x_i, d_i\}|a, b, \Theta) &= p(\{x_i, d_i\}|a, b, \Theta) \\ &= p(x_i, d_i \in pre|a, b, \Theta) \\ &\quad + p(x_i, d_i \in STN|a, b, \Theta) \\ &\quad + p(x_i, d_i \in post|a, b, \Theta) \end{aligned} \tag{7}$$

By expanding the probabilities in Eq. 7 using the Bayes' theorem, we get

$$\begin{aligned} L(\{x_i, d_i\}|a, b, \Theta) &= p(x_i|d_i \in pre, \Theta) \cdot p(d_i \in pre|a, b, \Theta) \\ &\quad + p(x_i|d_i \in STN, \Theta) \cdot p(d_i \in STN|a, b, \Theta) \\ &\quad + p(x_i|d_i \in post, \Theta) \cdot p(d_i \in post|a, b, \Theta) \end{aligned} \tag{8}$$

where the probability $p(x_i|d_i \in pre, \Theta)$ represents the emission probability in state *pre* and is computed using the standard probability density function of the log-normal distribution in the area before STN:

$$p(x_i, pre|\Theta) = \frac{1}{x_i \hat{\sigma}_{pre} \sqrt{2\pi}} \exp -\frac{(\ln(x_i) - \hat{\mu}_{pre})^2}{2\hat{\sigma}_{pre}^2}, \tag{9}$$

using parameters of the log-normal distribution $\hat{\mu}_{pre}$ and $\hat{\sigma}_{pre}$, obtained in the training phase according to Eqs. 1 and 2 respectively. The probabilities $p(x_i|STN, \Theta)$ and $p(x_i|post, \Theta)$ for NRMS distribution inside and beyond the STN are computed accordingly. The class membership probabilities $p(pre|a, b, \Theta)$ from Eq. 8 (similarly for states *STN* and *post*) depend on the distance between depth d_i and currently assumed STN borders a and b and are computed from the sigmoid transition functions as follows:

$$p(d_i \in pre|a, b, \boldsymbol{\Theta}) = (1 - S_{en}(d_i, a|\boldsymbol{\Theta}))/z_i$$
$$p(d_i \in STN|a, b, \boldsymbol{\Theta}) = S_{en}(d_i, a|\boldsymbol{\Theta}) \cdot S_{ex}(d_i, b|\boldsymbol{\Theta})/z_i \qquad (10)$$
$$p(d_i \in post|a, b, \boldsymbol{\Theta}) = (1 - S_{ex}(d_i, b|\boldsymbol{\Theta}))/z_i$$

using the sigmoid transition functions S_{en} and S_{ex}:

$$S_{en}(d_i) = \left(1 + \exp -(\beta_{en}^0 + \beta_{en}^1(a - d_i))\right)^{-1} \qquad (11)$$

for STN entry and equivalently

$$S_{ex}(d_i) = \left(1 + \exp -(\beta_{ex}^0 + \beta_{ex}^1(b - d_i))\right)^{-1} \qquad (12)$$

for STN exit. The z_i in Eq. 10 is a normalization coefficient ensuring that the class membership probabilities add to one under all circumstances[3]:

$$z_i = (1 - S_{en}(d_i, a|\boldsymbol{\Theta})) + S_{en}(d_i, a|\boldsymbol{\Theta}) \cdot S_{ex}(d_i, b|\boldsymbol{\Theta}) + (1 - S_{ex}(d_i, b|\boldsymbol{\Theta})). \qquad (13)$$

In case of the extended model $flex2$, the minimization will take the following form:

$$\{a, b\} = \underset{a,b}{arg\min} \left[\sum_{i=1}^{N} (-\ln(L(d_i, a, b|\boldsymbol{\Theta})) - \lambda ln(p_a(a|\boldsymbol{\Theta}') \cdot p_b(b|\boldsymbol{\Theta}'))) \right] \qquad (14)$$

where the summation $L(x_i, a, b|\boldsymbol{\Theta})$ is the same as in Eq. (6) and the new $p_a(a|\boldsymbol{\Theta}')$ and $p_b(b|\boldsymbol{\Theta}')$ are probabilities of STN entry at depth a and exit at depth b, computed from the normal probability density function

$$p_a(a|\boldsymbol{\Theta}') = \frac{1}{\sigma_a \sqrt{2\pi}} \exp -\frac{(a - \mu_a)^2}{2\sigma_a^2} \qquad (15)$$

and represent the probability of STN entry at depth a and exit at depth b. The parameter λ can be used to assign more/less importance to the a-priori depth distribution, compared to the observation-based likelihood element. In case of the presented results, we set the value of $\lambda = 1.75$ which optimized train-set accuracy.

As this process can be vectorized and the parametric space is only two-dimensional and bounded, standard optimization algorithms with empirical gradient can be used to search for optimal parameters. In our case, we used constrained optimization with conditions requiring that $a < b$ (the entry depth a is lower or equal to exit depth b), $a \geq d_1$ and $b \leq d_N$ (entry and exit depths must be in the range of the data).

The parametric space may contain local optima (depending on the shape of NRMS values along given trajectory) and it is therefore very useful to provide

---

[3] Value of this normalization coefficient will however be close to one in most circumstances and reaches around 1.2 in the extreme case when $a = b$ using sigmoid parameters from Fig. 1.

reasonable initialization of a and b. In our implementation, the initialization was set as the mean entry and exit depths from the training data: μ_a and μ_b[4]. Note that both a and b are real numbers and are not restricted to the set of actually measured depths.

2.4   Crossvalidation

To evaluate the proposed model on real data and compare its classification ability against existing models, we evaluated the model in a 20-fold crossvalidation: in each fold, 5 % of available trajectories were left out for validation, while the remaining data were used for estimation of model parameters. This lead to 20 sets of error measures for each classifier which were than averaged to obtain final estimates. Larger number of crossvalidation folds was chosen in order to obtain better estimate of error variability on different validation datasets.

The models compared were (i) Bayes classifier from [9] based on discrete joint probability distribution of NRMS and depth and an (ii) HMM model, based on the same discrete probability distribution (used as emission probabilities), with transition probabilities estimated from the training data in a standard way and two variants of the proposed model: (iii) *flex1*, based solely on NRMS and (iv) *flex2* with distribution of entry and exit depths.

3   Experimental Results

3.1   Data Summary

In total, we collected 6576 signals from 260 electrode passes in 117 DBS trajectories in 61 patients. Length of recorded signals was 10 s. After discarding non-stationary signal segments, the mean length of raw signal segment that entered the NRMS calculation was 8.76 s (median 9.67 s). In each crossvalidation fold, 13 electrode passes were used for validation, while the remaining 247 were used for training.

3.2   Classification Results and Discussion

Mean values of classification sensitivity, specificity and accuracy are presented in Table 1, while distribution of these error measures on the 20 validation sets can be found in Fig. 3. Even though the results of all methods were very similar (as can be seen especially in Fig. 3), the highest mean test accuracy was achieved by the *hmm* model – 90.2 %, closely followed by the *flex2* model with 90.0 %. Both models were also best in terms of specificity, while the best validation set sensitivity was achieved by the *hmm* and *bayes* classifiers.

Comparing two variants of the proposed method, the *flex2* model with entry depth distribution achieved better results than the NRMS-only variant *flex1*. The latter model tended to converge to local optima on trajectories with high noise level or non-standard NRMS shape.

---

[4] In the case with no entry/exit depth distribution, the initial parameters were set as the middle of the trajectory for a and the 3/4 of the trajectory for b.

Table 1. Classification results (error measures from the 20-fold crossvalidation) comparing the results of Bayes classifier [9] (*bayes*), Hidden Markov model (*hmm*), suggested model based solely on the NRMS (*flex1*) and extended model with distribution of STN entry and exit depth (*flex2*). See also Fig. 3.

	Train			Test		
	Accuracy	Sensitivity	Specificity	Accuracy	Sensitivity	Specificity
bayes	90.4	84.1	94.1	89.0	82.5	92.8
hmm	91.3	83.8	95.7	90.2	83.1	94.3
flex1	88.5	80.9	92.9	88.0	80.6	92.2
flex2	90.1	83.2	94.1	90.0	83.1	94.1

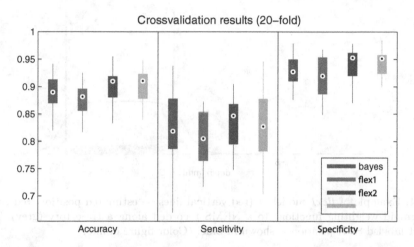

Fig. 3. Classification results on the 20 validation sets: *bayes* classifier [9], Hidden Markov model (*hmm*), suggested model, based exclusively on NRMS (*flex1*) and extended model with added a-priori entry and exit depth distribution (*flex2*).

3.3 Fitting of Individual Trajectories and Log-Likelihood Function Shape

Apart from the overall results, we also evaluated results on individual trajectories. The *bayes* model, which from definition put no constraints on the resulting label vector, was capable of classifying non-consecutive trajectories (interrupted STN labels) — this may have lead to the rather high sensitivity on the training data. As for the proposed models, the *flex1* NRMS-only variant tended to fit zero-length STN near the end of the trajectory in cases of non-standard STN passes where the NRMS did not exhibit the standard low–high–low profile or contained strong local peaks. The addition of entry and exit depth distribution in the *flex2* model variant reduced this problem and lead to improved classification accuracy.

An example of a successful STN classification on a typical trajectory using the *flex1* model can be seen in Fig. 4, while the corresponding negative log-likelihood function from Eq. 8 can be seen in Fig. 5. Note that the log-likelihood function is defined only for $a \leq b$. In the case of the *flex2* model, the values of the likelihood function around the a-priori expected entry and exit depth are further reduced by the additional component in Eq. 14, which increases the performance especially in cases with high noise in NRMS values.

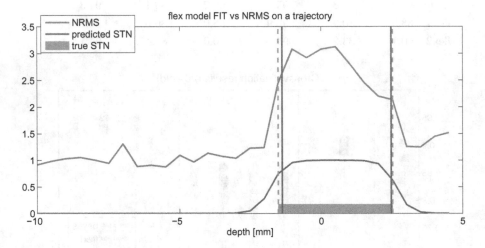

Fig. 4. Example of *flex1* model fit (red vertical lines — estimated position, red curve — sigmoid weighting function) to a NRMS recorded along a trajectory (grey). The expert-labeled STN position is shown in blue. (Color figure online)

4   Discussion and Further Work

The presented model achieved comparable accuracy to existing approaches, represented by bayesian classifiers [9] and HMM [14]. The results of HMM and hidden semi-markov models, presented by Taghva et al. [13] were much superior, but were evaluated on simulated data only. In summary, the presented extended model (*flex2*) achieved mean classification accuracy 90.0 %, sensitivity 83.1 % and specificity 94.1 % on the test set. As seen from the heavy overlap of different method's results, clearly visible in Fig. 3, we can conclude that it is rather robustness of the NRMS feature itself than the model structure, that has major impact on the results.

The main aim of this paper was to prove feasibility and efficacy of a probabilistic model which is variable in structure and can potentially be used for fitting of an anatomical 3D model to μEEG signals in multi-electrode setting. In such case, the inside and outside volume of the anatomical model would yield different emission probability distribution and further constraints or penalization

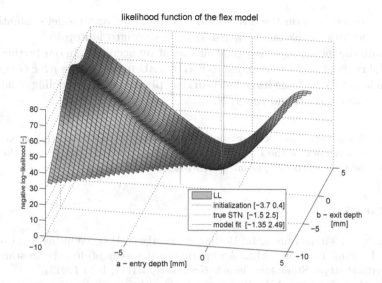

Fig. 5. Negative log-likelihood function of the *flex1* model shown as a function of hypothetical STN entry (*a*) and exit (*b*) depth. The vertical lines show initialization (magenta), model fit (red) and expert labels (blue). (Color figure online)

on model shift, scaling or rotation could be added easily into the minimization function. We have shown, that such addition of further constituents — such as the entry and exit depth in case of the *flex2* model — can be done and can contribute to improved classification accuracy.

The key part of the presented model is the use of smooth state transition functions, which ensure smooth shape of the resulting likelihood function and enable the use of general-purpose optimization techniques for model fitting. Another consequence of the use of sigmoid transition functions is that the detected transition point does not have to be truncated to a position of available measurement, but can be at an arbitrary position between states (i.e. the detected entry and exit depths are real numbers, not constrained by the depths where μEEG recordings are available).

The drawbacks of the presented model are that contrary to Bayes classifier or an HMM it is not straightforward to convert the presented method to an online algorithm, used e.g. during the surgery. Another weak point is the lack of closed-form solution to model evaluation and the necessity to use general optimization. Thanks to the low dimension[5] and small size of the parametric space, this does not pose a real problem in the presented settings, as the parameter estimation

---

[5] Dimension of the parametric space searched during the evaluation phase is two, due to two optimized parameters: STN entry *a* and exit *b*, both in the range of recorded depths. The search space is further reduced by the conditions defined at the end of Model Evaluation section, especially $a \leq b$.

took on average 0.9 s on the 247 training trajectories and model evaluation on all 260 trajectories took on average 4.5 s on a standard laptop PC.

Overall, the model provided good classification accuracy. In our further work, the model concept will be extended to fitting a 3D model to the μEEG trajectories, which may bring benefits to both surgical planning and modeling of neuronal activity within and around the STN.

Acknowledgement. The work presented in this paper has been supported by the students' grant agency of the CTU, no. SGS16/231/OHK3/3T/13, and by the Grant Agency of the Czech republic, grant no. 16-13323S.

References

1. Abosch, A., Timmermann, L., Bartley, S., Rietkerk, H.G., Whiting, D., Connolly, P.J., Lanctin, D., Hariz, M.I.: An international survey of deep brain stimulation procedural steps. Stereotact. Funct. Neurosurg. **91**(1), 1–11 (2013)
2. Aboy, M., Falkenberg, J.H.: An automatic algorithm for stationary segmentation of extracellular microelectrode recordings. Med. Biol. Eng. Comput. **44**(6), 511–515 (2006). http://www.ncbi.nlm.nih.gov/pubmed/16937202
3. Bakstein, E., Schneider, J., Sieger, T., Novak, D., Wild, J., Jech, R.: Supervised segmentation of microelectrode recording artifacts using power spectral density. In: 2015 37th Annual International Conference of the IEEE Engineering in Medicine and Biology Society (EMBC), vol. 2015-Novem, pp. 1524–1527. IEEE, August 2015. http://ieeexplore.ieee.org/lpdocs/epic03/wrapper.htm?arnumber=7318661
4. Benabid, A.L., Pollak, P., Gao, D., Hoffmann, D., Limousin, P., Gay, E., Payen, I., Benazzouz, A.: Chronic electrical stimulation of the ventralisintermedius nucleus of the thalamus as a treatment of movement disorders. J. Neurosurg. **84**(2), 203–214 (1996). http://dx.doi.org/10.3171/jns.1996.84.2.0203
5. Cagnan, H., Dolan, K., He, X., Contarino, M.F., Schuurman, R., van den Munckhof, P., Wadman, W.J., Bour, L., Martens, H.C.F.: Automatic subthalamic nucleus detection from microelectrode recordings based on noise level and neuronal activity. J. Neural. Eng. **8**(4), 46006 (2011). http://www.ncbi.nlm.nih.gov/pubmed/21628771, http://dx.doi.org/10.1088/1741-2560/8/4/046006
6. Gross, R.E., Krack, P., Rodriguez-Oroz, M.C., Rezai, A.R., Benabid, A.L.: Electrophysiological mapping for the implantation of deep brain stimulators for Parkinson's disease and tremor. Mov. Disord. **21**(Suppl. 1), S259–S283 (2006). http://dx.doi.org/10.1002/mds.20960
7. Guillen, P., Martinez-de Pison, F., Sanchez, R., Argaez, M., Velazquez, L.: Characterization of subcortical structures during deep brain stimulation utilizing support vector machines. In: 2011 Annual International Conference of the IEEE Engineering in Medicine and Biology Society, vol. 6, pp. 7949–7952. IEEE, August 2011. http://ieeexplore.ieee.org/xpls/absall.jsp?arnumber=6091960, http://ieeexplore.ieee.org/lpdocs/epic03/wrapper.htm?arnumber=6091960
8. Hammerla, N.Y., Plötz, T.: Let's (not) stick together: pairwise similarity biases cross-validation in activity recognition. In: Proceedings of the 2015 ACM International Joint Conference on Pervasive and Ubiquitous Computing, pp. 1041–1051 (2015)

9. Moran, A., Bar-Gad, I., Bergman, H., Israel, Z.: Real-time refinement of subthalamic nucleus targeting using Bayesian decision-making on the root meansquare measure. Mov. Disord. **21**(9), 1425–1431 (2006). http://www.ncbi.nlm.nih.gov/pubmed/16763982, http://dx.doi.org/10.1002/mds.20995

10. Novak, P., Daniluk, S., Ellias, S.A., Nazzaro, J.M.: Detection of the subthalamic nucleus in microelectrographic recordings in Parkinson disease using the high-frequency (> 500 hz) neuronal background. J. Neurosurg. **106**(1), 175–179 (2007). http://dx.doi.org/10.3171/jns.2007.106.1.175

11. Novak, P., Przybyszewski, A.W., Barborica, A., Ravin, P., Margolin, L., Pilitsis, J.G.: Localization of the subthalamic nucleus in Parkinson disease using multiunit activity. J. Neurol. Sci. **310**(1–2), 44–49 (2011). http://linkinghub.elsevier.com/retrieve/pii/S0022510X11004448

12. Shamir, R.R., Zaidel, A., Joskowicz, L., Bergman, H., Israel, Z.: Microelectrode recording duration and spatial density constraints for automatic targeting of the subthalamic nucleus. Stereotact. Funct. Neurosurg. **90**(5), 325–334 (2012). http://www.ncbi.nlm.nih.gov/pubmed/22854414, http://www.karger.com/doi/10.1159/000338252

13. Taghva, A.: Hidden Semi-Markov Models in the computerized decoding of microelectrode recording data for deep brain stimulator placement. World Neurosurg. **75**(5-6), 758–763.e4 (2011). http://www.ncbi.nlm.nih.gov/pubmed/21704949, http://linkinghub.elsevier.com/retrieve/pii/S187887501000848X

14. Zaidel, A., Spivak, A., Shpigelman, L., Bergman, H., Israel, Z.: Delimiting subterritories of the human subthalamic nucleus by means of microelectrode recordings and a Hidden Markov Model. Mov. Disord. **24**(12), 1785–1793 (2009). http://www.ncbi.nlm.nih.gov/pubmed/19533755, http://dx.doi.org/10.1002/mds.22674

Social Networks and Process Analysis in Biomedicine

Multidisciplinary Team Meetings - A Literature Based Process Analysis

Oliver Krauss$^{(\boxtimes)}$, Martina Angermaier, and Emmanuel Helm

University of Applied Sciences Upper Austria, 4232 Hagenberg, Austria
{Oliver.Krauss,Martina.Angermaier,Emmanuel.Helm}@fh-hagenberg.at
https://www.fh-ooe.at/en/

Abstract. Multidisciplinary Team Meetings (MDTM) are conducted
to discuss the treatment of one or more patients. This paper discusses
MDTM with a focus on tumor treatment and shows workflows in different
settings, identifies organizational and technical problems in the MDTM
and solutions thereof. It aims to answer the following research questions:
(RQ1) *What is the current state of the art in MDTM?* (RQ2) *How are
they conducted and what is the variation in different hospital settings?*
(RQ3) *What technical problems and possible solutions thereof exist?* This
is done by conducting a literature review entailing a forward search of 837
papers and a backward search. The results show that a unified workflow
model for MDTM can't be found since they are highly dependent on
institutional and tumor dependent specifics. The identified problems and
solutions show a lack of research towards technical solutions and process
interoperability. An outlook on extending research in these areas is given.

Keywords: Tumor board · Multidisciplinary team meeting · Workflow ·
Process analysis

1 Introduction

Multidisciplinary Team Meetings (MDTM), also called Tumor Boards (TB),
are meetings between medical practitioners of different medical disciplines to
determine the diagnosis and further treatment of one or more patients. The
MDTM can either be carried out locally or distributed via telecommunication
solutions [1]. While MDTMs are mostly associated with tumor treatment other
treatment sectors have been known to use them as well, such as psychiatric care
[2].

The MDTM has not been proven to significantly improve patient care. Wright
et al. [3] and Patkar et al. [23] did a systematic review and both highlight the
absence of good quality empirical evidence. However, MDTMs are well accepted
and the lack of evidence may not be interpreted as a proof of ineffectiveness
[23]. Moreover most experts and national and international guidelines demand
MDTMs in oncology [22]. For example, according to the Austrian cancer program every patient must have access to a TB [4]. Additionally insufficient communication and missing information are among the major factors contributing

© Springer International Publishing Switzerland 2016
M.E. Renda et al. (Eds.): ITBAM 2016, LNCS 9832, pp. 115–129, 2016.
DOI: 10.1007/978-3-319-43949-5_8

to adverse events in medicine (i.e. unintended injuries caused by medical management rather than the disease process) [30].

Guidelines on the MDTM define participants in the board, as well as the process of the board itself and medical information that is needed for it to function [3,5]. The Integrating the Healthcare Enterprise (IHE) Profile Cross-enterprise Tumor Board Workflow Definition (XTB-WD) identifies the workflow of a MDTM in the oncological process [6]. XTB-WD as well as other literature on MDTM identifies several problems concerning the organizational and technical parts of those meetings, as well as in the necessary preparation and documentation after them [6]. Other works show that the MDTM is time-consuming for practitioners that participate in it [7,8].

This paper shows MDTM workflows in single institution as well as distributed and cross-enterprise settings to answer the questions *what is the current state of the art in MDTM (RQ1)* and *how are they conducted and what is the variation in different hospital settings (RQ2)?* Problems, areas for improvement and possibilities to reduce organizational overhead in the workflow will be identified. Solutions specifically in the areas of workflow assistance and automation as well as healthcare interoperability will be suggested to set a basis for further research and show *what technical problems and possible solutions thereof exist (RQ3)?*

The remainder of this paper describes the methods (Sect. 2) of identifying the workflow and problems using a literature research, presenting relevant results (Sect. 3) and a discussion (Sect. 4) that gives possible proposals for solutions of those areas as well as showing further areas of research.

2   Methods

The findings in this paper were achieved through a research of current literature with searches of scientific databases conducted on 2. Feb 2016.

Search Strategy - In a literature based search, relevant scientific articles were identified by searching the following scientific databases: Science Direct, PubMed, Association for Computer Machinery (ACM), Institute of Electrical and Electronics Engineers (IEEE). The terms (1) tumor board OR (2) multidisciplinary team meeting OR (3) multidisciplinary medical team meeting were used. Searches were limited to Abstract, Title and Keywords and restricted to the year 1990 and newer. In addition after the forward search a backward reference search on the selected articles was performed.

Inclusion Criteria - Since several definitions of a TB exist (MDTM or tumor conference), articles were selected by applying to the following definition of a TB: "A meeting in person or over teleconferencing solutions between medical personnel of different disciplines to discuss the medical situation of one or more patients and further treatment thereof". Articles had to either (1) discuss how a MDTM was conducted, including required pre- and post- board activities, (2) show specific technical or organizational problems or improvement measures for MDTM or (3) define guidelines on MDTMs.

Exclusion Criteria - (1) Any Publication not written in English or where the full-text was not available was excluded. Furthermore (2) due to a significant lack of publications on MDTM outside of the context of tumor treatment only publications on that topic were selected. (3) Publications only discussing quality, cost, feasibility or impact on patient care were excluded as well. Li et al. [1] was excluded from selection since Robertson et al. [7] was conducted in the same setting 2 years later by similar authors with similar, but updated findings, thus only Robertson et al. [7] will be included in the results so findings are not duplicated. Other papers were partially excluded if the full-text analysis showed that the same data base, questionnaires or interviews, were used to conclude specific findings. Kane et al. [29] and Kane et al. [27] are both based on the same larger on-going study, thus they were partially seen as one.

Selection of Articles - Articles were pre-filtered for eligibility by screening the abstract and keywords. After pre-filtering a full-text analysis of the remaining articles was performed based on the inclusion- and exclusion criteria.

3   Results

Of the initially found 837 articles, 25 met the inclusion criteria. Table 1 shows a more detailed look in the selection process. After selection the final articles were categorized according to the following criteria what information they provided: (1) participants of a MDTM, (2) patient participated in MDTM, (3) information required for a MDTM, (4) workflow of MDTM, (5) problems identified, (6) solutions identified. Articles may be in one or more categories as can be seen in Table 2. The findings will be discussed per category.

3.1   Participants in a MDTM

Table 3 lists the key participants which were identified by the literature review. The specialties of oncologists, nurses and other specialists depend on the type of the MDTM. All 15 articles which show participants mention oncologists as key members of the MDTM. Pathologists and radiologists are mentioned by 14 articles and surgeons by 13. Two third of the papers list radiotherapists as important participants.

Table 1. Amount of Articles in selection process of scientific publications

Step in literature search	Amount of Articles
Search in databases	837
After duplicates removed from pool	728
Remaining articles after abstract selection	82
Remaining articles after full text analysis	19
Articles after searching referenced articles	25

Table 2. Categorization of articles

Category	Articles	Amount
Shows participants	[3, 7, 9, 10, 13, 15, 16, 19–22, 24, 26–28]	15
Patient participates	[5, 9, 10, 25]	4
Identifies information required	[5, 7, 9, 10, 15, 17, 19, 20, 24, 26, 28, 29]	12
Shows workflow	[3, 7, 12, 13, 16–21, 26–28]	13
Identifies problem(s)	[3, 7, 8, 10–15, 19, 23, 25–29]	16
Identifies solution to problem(s)	[3, 7, 10–13, 15, 16, 18, 20, 22, 25–29]	16

Table 3. Participants in a MDTM

Participant	Articles	Amount
Oncologists (different specialties)	[3, 7, 9, 10, 13, 15, 16, 19–22, 24, 26–28]	15
Pathologists	[3, 7, 9, 13, 15, 16, 19–22, 24, 26–28]	14
Radiologist	[3, 7, 9, 10, 13, 15, 16, 19, 21, 22, 24, 26–28]	14
Surgeons	[3, 7, 9, 13, 15, 16, 19, 20, 22, 24, 26–28]	13
Radiotherapists	[3, 7, 9, 13, 16, 19–21, 26, 27]	10
Nurses (different specialties)	[3, 7, 15, 22, 24, 26, 27]	7
Coordinator; administrator	[3, 19, 22, 24, 26, 27]	6
Researcher	[15, 21, 26–28]	5
Junior physicians; trainees; students	[16, 21, 22, 27, 28]	5
Other specialists (different specialties)	[10, 15, 19, 20, 22]	4
Dietitians	[3, 15, 22, 27]	4
Physiotherapists	[3, 22, 26, 27]	4
Psychologists	[3, 7, 9]	3
Social workers	[3, 7, 22]	3
Data mangers	[7, 26, 27]	3
General practitioners	[3, 7, 27]	3
Palliative care physicians	[3, 22]	2
Occupational therapists	[3, 22]	2
Pastoral care	[3]	1
Medical geneticists	[21]	1
Speech and language therapist	[27]	1
Anaesthetists	[15]	1

Jalil et al. [15] notes that anaesthetists are usually not included in a MDTM but that they should be since the anaesthetist may later cancel surgeries due to a lack of investigation.

3.2 Patient Participates

Tőkés, T., Torgyík et al. [9] shows that communication with the patient is important in a MDTM, since a patient may prefer treatments based on the impact on the patient's life such as quality of life or preserving the body image. Huber et al.

[10] shows that patient involvement in a tumor board is highly valued by patients as well as physicians, although it also involves an increased workload to the physicians. The guidelines defined in Deneuve et al. [5] allow the patient to note his interest in participating in the MDTM but also note that patient participation is debatable. Lamb et al. [25] states that there are not many studies addressing the patient's involvement. However nurses are perceived to represent patient's preferences.

3.3   Information Required in Advance of a MDTM

Table 4 summarizes the information which is required in advance of a MDTM. 12 articles contain required information for a MDTM. Imaging results such as x-ray images, CT-scans or MRI images are listed most frequently. Information about the patient's medical history and the earlier treatments, for example as a patient summary is mentioned by 8 out of 12 articles. Further important information is histological findings or laboratory results which are mentioned by more than half of the articles.

3.4   Workflows

Thirteen different workflows for MDTM were identified (see Table 2). Note that not all sources described the workflow directly but could be inferred through other means such as guidelines, role and activity descriptions. From the identified workflows it was attempted to create a unified workflow model that shows the tumor board process. However this was not possible since most sources show different workflows, do not show parts of the workflow, show different orders of the same tasks or have different roles process the tasks.

The identified workflows were compared by splitting them into three separate parts dependent on when they happen. First comes the preparation work that needs to be done before a MDTM can even be executed, and is shown in Fig. 1. Secondly the MDTM and the therein entailed process steps are shown in Fig. 2. Lastly Fig. 3 entails the steps that need to be taken after a MDTM was conducted such as documentation work. It is important to note that the different authors used different terms, and it was sometimes unclear if they meant the same thing or not (ex. Jazieh et al. [16] mentions *Prepare image films*. It is unknown if they meant radiology or pathology images or both). Where a definitive identification was not possible the steps are listed separately in the images. Some of the roles, such as participants, lead, moderator etc. in Figs. 1, 2 and 3 are not shown in Table 3 since those roles are taken on only during the MDTM such as the participants being all members of the MDTM.

Though neither of the three images correspond to a valid process model, parts of the Business Process Model Notation (BPMN) were taken to make the processes understandable. On the left side of Figs. 1, 2 and 3 the "Generic" Process steps can be seen. They summarize the tasks shown on the right side and show which papers cite steps. The right side shows the actual steps as described in literature, wherein a step has the executing role on top, and the

Table 4. Information required in advance of a MDTM

Category	Articles	Amount
Imaging results (X-ray, CT-scan, MRI)	[5,7,9,10,15,17,19,24,28,29]	10
A summary of earlier treatments; patient summary; medical history	[7,9,15,17,19,20,24,26]	8
Histological findings; laboratory results	[5,7,9,15,20,24,26]	7
TNM cancer stage, Clinical staging	[5,19,20,26]	4
Psychological risk factors	[7,10,15,24]	4
Co-morbidities	[15,17,20,24]	4
General health status (e.g. ECOG)	[5,17,19,20]	4
Patient demographic factors (age, sex, income, education level)	[7,9]	2
Symptoms	[19,20]	2
Patient treatment guidelines	[19,20]	2
Related journal articles, reviewed literature	[19,20]	2
Patient's view or preferences	[20,24]	2
Pathology images	[28,29]	2
Opinion of family physician	[5]	1
Epidemiological and social data	[5]	1
Diagnosis	[20]	1
Tumor marker value	[20]	1
Question(s) to the tumor board	[20]	1
The referring consultant	[26]	1
Names of pathologists and radiologists who contributed the findings	[26]	1
Differential diagnosis; what was ruled out	[26]	1

actions taken below, or alternatively has only the actions taken shown if no roles were defined. Process Flows are shown with arrows in the flow direction. If a flow is shown horizontally or vertically has no meaning and was simply done in each case to save space. Process steps without flows do exist and indicate that they were the only step mentioned in the referenced paper. Steps where no order was defined in the literature are connected with a "+" symbol meaning that any order is possible but they are still discussed sequentially. An "x" symbol means that only one path will be followed. All steps require user interaction if not mentioned otherwise, indicated by a gearwheel in the left upper corner of the step.

Figure 1 shows the preparatory steps for a MDTM. The general workflow shows that first patients are selected for a tumor board and a list of patients will be compiled. Afterwards the patient data for the selected patients is prepared or generated. The MDTM is organized, times/places are selected and attendees are invited. The data to be discussed in the MDTM is sometimes made available to the participants beforehand. Especially the point prepare patient data shows many (7) different paths for how data can actually be acquired. This may be explained by the hospitals varying use of healthcare interoperability standards and technical solutions/platforms in their hospital environments.

The steps that will be taken during the MDTM are shown in Fig. 2. First the meeting is opened and the agenda is discussed. All following steps are done per patient case, starting with the presentation of the case and review of relevant medical data. The case is then discussed by the group, sometimes according to guidelines and treatment plans. Finally either the group either decides that further information is required for selection of an appropriate treatment, or decides on a treatment plan and documents this fact. It is notable that automatically take records and statistics is the only step identified that requires no user interaction.

Figure 3 shows the steps that need to be taken after a MDTM was conducted. Firstly the results of the MDTM are documented and the documentation is afterwards spread either to the patients themselves, the treating physicians or tumor board participants. The steps validate and analyse were the only ones which could not be identified in the general workflow since no mention is made of when they occur. It can be assumed however that documented results should be validated before spreading them Kane et al. [26] does not mention the spreading of the information. Note that both however will occur after the documentation of results, which in Kane et al. [26] as well as Mangesius et al. [13] is done during MDTM (see Fig. 2) instead of after.

3.5   Identified Problems

Robertson et al. [7] describes a hospital using PowerPoint for prepared radiology and pathology images. They describe a high (up to 12 h) time of preparation for those images. Jalil et al. [15], Lamb et al. [25] and Kane et al. [29] identify time pressure and a lack of preparation of cases as a negative influence on the MDTM as well. Frykholm et al. [17] also notes that IT-support is not accepted if it adds work for the participants. Look Hong et al. [8] identifies a "Value for Time Balance" as a significant factor in MDTM. Specifically a lack of prior reviews of cases and efficient management were identified as hindrances for MDTM. Munro et al. [12] notes that virtual MDTM are more time-consuming than regular ones. Patkar et al. [23] discusses the assessment of the MDT performance or find the balance of educational and care delivery objectives.

In Frykholm et al. [17] medical personnel requested more information and an effective to be provided, showing a lack of available data. Jalil et al. [15] identifies lack of available information as a negative factor. Mostly missing patient history, investigation results but also co-morbidities, psychological factors and the patient's wishes are sometimes unavailable but necessary in the MDTM. Avila-Garcia et al. [28] shows that during the MDTM the discussion about the images may be limited, since the each member interested in viewing a specific image or structure waits to ask and direct the radiologist to show the image

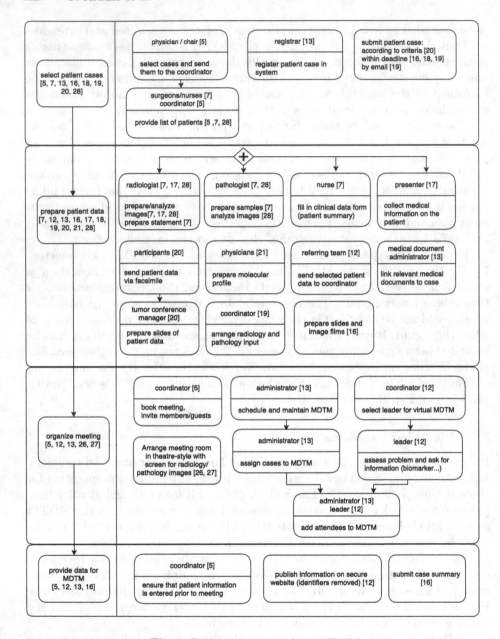

Fig. 1. Preparation steps for a MDTM

and region of interest, potentially diverting the attention away from the main discussion.

Huber et al. [10], Choy et al. [11] and Patkar et al. [23] identify the lack of patient involvement as a serious problem because treatment cannot be recommended based

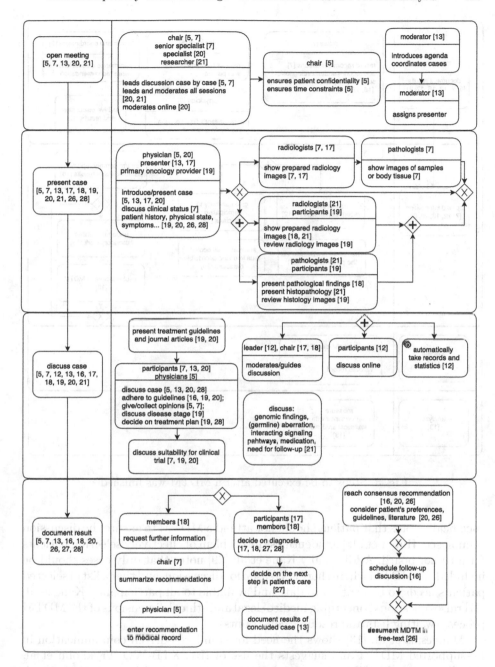

Fig. 2. Steps during the conduction of a MDTM

on the patients individual needs. [12] mentions that it is easier to involve the patient in a virtual MDTM as opposed to a face-to-face one. To the contrary Wright et al. [3]

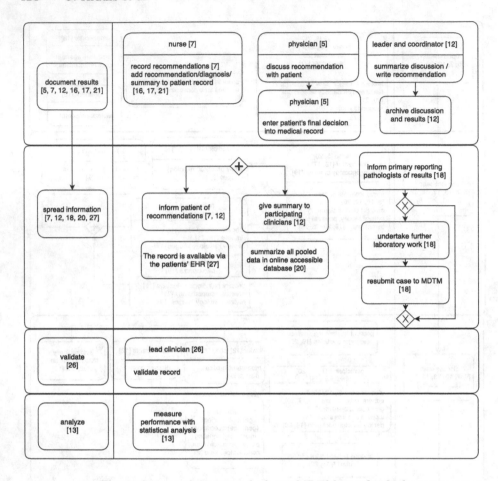

Fig. 3. Steps to be executed after a MDTM was finished

specifically notes that patient should not attend a MDTM to ensure the discussion is unbiased. Huber et al. [10] identifies the problem of effectively selecting patients for a tumor board. Additionally Kane et al. [29] notes that only cases should be included in a MDTM where there is a reason for discussion, and that a list of selected patients needs to be kept up to date and available to all participants. Kane et al. [27] raises questions concerning medico–legal and ethical dimensions of the MDTM process (waiting lists and resource constraints).

Mangesius et al. [13] shows the need for standards-driven communication in IT-supported MDTM and suggests the use of IHE XTB-WD. Frykholm et al. [17] discusses a MDTM where an electronic health record is available for each patient. It is rarely used since it takes a long time to access the information. A survey on their IT-solution used in MDTM also showed that participants asked for keywords on important medical information of the patient to further reduce time spent. Kane et al. [29] cites a lack of interoperability when providing information to local practitioners as well as automatically inputting information into National Data Repositories.

Jalil et al. [15] identifies missing members of the MDTM as a problem. Kane et al. [29] also reports problems in attendance of members. An identified reason for a lack of attendance was the problem of finding appropriate timeslots fitting for all members.

Stevenson et al. [19], Kane et al. [29] describe technical challenges especially connection problems for web conferences. The members of the MDTM in Robertson et al. [7] use laser-pointers to discuss the images, which cannot be seen by the participants so one member also moves the mouse pointer to the same location. A lag while transmitting images was also identified as a major technical problem. Sallnäs et al. [14] shows that accurate pointing, specifically at medical images, is important for clarity in a MDTM discussion.

Patkar et al. [23] requires that the MDT recommendations adhere with guidelines and that they are followed in the practice.

Patkar et al. [23] shows that ensuring consistency of the patient data can be a challenge for the data collection. In addition medicolegal liability should be considered. Similarly Kane et al. [26] describes the challenge of record-keeping - that the right information is recorded in the right way without taking too much time. They found that audio-visual recording was not accepted by the professionals and a well-structured form became too complex. Kane et al. [27] also elaborates on record-keeping and emphasizes the need for high data quality in the record. The identified problems range from record availability over the need for more patient data to governance issues (i.e. who is responsible for a specific record). Kane et al. [29] also notes that it needs to be documented who attended the MDTM.

3.6 Identified Solutions

To improve patient involvement Huber et al. [10] suggests an interdisciplinary counseling service that is essentially an individual MDTM for that patient. A study conducted proves that this method is highly rated by both physicians as well as patients since it improves patient-physician communication and gives the patient insights into the reasoning behind their treatment plans [10]. In a similar approach Choy et al. [11] conducted a study in which patients attended the regular TB with mostly positive feedback from patients and physicians. Jalil et al. [15] also recommends involving the patient in the MDTM to improve decision making, but notes that some physicians argue against implementing this.

Wright et al. [3], Mangesius et al. [13] and Frykholm et al. [17] identify roles with specific responsibilities, such as a meeting coordinator, as a way to guide the format of a MDTM. Wright et al. [3] also identifies the need for a written protocol that describes members of the MDTM, the meeting format, responsibilities and roles as well as guidelines on maintaining patient confidentiality. Jalil et al. [15] similarly suggests appointing effective chairing and refining criteria to improve effectiveness of the MDTM.

Munro et al. [12] shows virtual MDTM as a solution for several organizational problems face-to-face MDTM have, such as scheduling conflicts or unavailable preparation time. A vMDTM allows the participants to prepare and discuss data as needed since it can be conducted over a longer time asynchronously. vMDTM also allow integration of the patient in the discussion if needed. El Saghir et al. [22] suggests tumor boards with a smaller group size if the health care resources are limited. In suboptimal settings the introduction of a video-conferencing tumor board counters limitations.

In Kane et al. [26] the record-keeping is based on a free text field and validated afterwards by the lead clinician. The MDTM members developed a shared language of abbreviations, combining their different jargons. Kane et al. [27] suggest reliable EHRs as a priority for MDTMs. This record should serve clinical audits and quality management functions and provide national repositories with information. However staff is reluctant to implement such a system due to anticipated extra time. In Kane et al. [29] the need for a formal record of proceedings is seen as the most important requirement.

Sallnäs et al. [14] uses laser pointers for accurate pointing during MDTM and the computer-mouse for IT-supported MDTM. In Avila-Garcia et al. [28] a horizontal multitouch-display, the DiamondTouch, was used to interact with the displayed images in a very intuitive and immediate way. Robertson et al. [7] suggests providing additional network capacity to prevent lag when transmitting images for a virtual MDTM.

Jazieh et al. [16] shows that a tumor board can be improved by using it as a learning platform for medical personnel in training, as well as a platform for cancer research. Chekerov et al. [20] proposes the presentation of guidelines and study result during tumor boards for continuous education.

Ashton [18] shows that selection of data for the MDTM is dependent on the patient case and can save time if it is known before preparation of data for the MDTM. Kane et al. [29] mentions that support of the preparation for MDTM should be prioritized.

Lamb et al. [25] proposes to strengthen the role of nurses because they focus on the patient's preferences, the co-morbidities and psychological and social issues. Furthermore adequate facilities and protected time for preparations have an impact on the quality.

4   Discussion

The results show that the MDTM workflows in each hospital are very similar (see Figs. 1, 2 and 3). They each have necessary preparation work, discuss patients one by one during the MDTM and the results are added to the patient act afterwards. All of the MDTM share similar tasks, such as preparing slides and images or adding documentation. However the results of combining 13 workflows show 49 differences in the details of their tasks. 35 step-orders are unique (meaning steps connected by flows or single steps which are unconnected), 4 of those orders can have different flows and in 10 tasks different roles are responsible for the task execution.

Hospitals can use interoperable solutions to a MDTM such as XTB-WD, since XTB-WD only identifies tasks on a level of Prepare and Finalize, leaving the minutia to the implementer. The identified tasks are similar to the ones observed in this publication. While this makes XTB-WD highly adaptive for different tumor boards this does not allow a partial automation or a definition of the board itself. However XTB-WD was intended to document a tumor board and these options are not an intended feature [6].

A distinct lack of technical solutions to the identified problems can be observed. More than half of the problems identified were of a technical nature or could be mitigated if not solved entirely by technical solutions or automation. For example a lack of preparation time, lack of available information or missing standards compliance could all be improved by technical solutions. However only a minority of solutions suggest a technical solution such as conducting vMDTM [2] or using EHR [27].

5   Conclusion and Outlook

Regarding the first research question (RQ1) *What is the current state of the art in MDTM?* 25 different articles were analyzed to answer this question. The answer seems to be that *MDTM are a well researched topic.* In conjunction the answer to (RQ2) *How are they conducted and what is the variation in different hospital settings?* shows *that MDTM have a core-workflow with highly adapted details depending on tumor-type and hospital environment it can be concluded that one single universal workflow cannot be defined.* This suggests that further research should focus on methods that allow the definition of workflows according to the needs of a specified MDTM.

To answer (RQ3) *What technical problems and possible solutions thereof exist?* several *different problems were identified. The lack of technical solutions indicates that there is still potential for research in these areas.* For example several tasks in the identified workflows could be automated, i.e. parts of the data-acquisition for the meeting preparation or documenting the tumor board. This could improve the efficiency of MDTM while also reducing the workload on medical personnel.

A lack of interoperability was also identified. Both from suggested solutions, and problems concerning MDTM conducted between different institutions, by acquiring data for a MDTM and when publishing data to other institutions (family physician). A mayor outcome of the literature research was that MDTM needs clearly specified language and common semantics to accurately describe a workflow for the MDTM (complications resulting from unclear semantics can be seen in Figs. 1, 2 and 3).

To verify the results, expert interviews were conducted with the heads of Austrian tumor competence centers in Linz. The interviewees highlighted the importance of codes and standards before, during and after MDTMs to enable technical integration and proper cooperation between the medical experts. They also stated that most of the preparation cannot be automated because physicians have to prepare the patients individually, understanding the case and its implications. Investigation concerning practitioner needs in tumor boards is necessary to understand the requirements correctly. Further expert interviews to validly determine the situation in Austria will be conducted and compared with the results herein.

References

1. Li, J., Robertson, T., Hansen, S., Mansfield, T., Kjeldskov, J.: Multidisciplinary medical team meetings: a field study of collaboration in health care. In: OZCHI 2008 Proceedings (2008)
2. Mohr, W.K.: A critical reappraisal of a social form in psychiatric care settings: the multidisciplinary team meeting as a paradigm case. Arch. Psychiatr. Nurs. **IX**(2), 85–91 (1995)
3. Wright, F.C., De Vito, C., Langer, B., Hunter, A.: The Expert Panel on Multidisciplinary Cancer Conference Standards, Multidisciplinary cancer conferences: a systematic review and development of practice standards. Eur. J. Cancer **43**, 1002–1010 (2007)
4. Mitglieder des Onkoligie-Beirates: Krebsrahmenprogramm Österreich. Bundesministerium für Gesundheit, Wien (2014)
5. Deneuve, S., Babin, E., Lacau-St-Guily, J., Baujat, B., Bensadoun, R.-J., Bozec, A., Chevalier, D., Choussy, O., Cuny, F., Fakhyr, N., Guigay, J., Makeieff, M., Merol, J.-C., Mouawad, F., Pavillet, J., Rebiere, C., Righini, C.-A., Sostras, M.-C., Tournaille, M., Vergez, S.: Guidelines(short version) of the French Otorhinolaryngology Head and Neck Surgery Society (SFORL) on patient pathway organization in ENT: the therapeutic decision-making process. Eur. Ann. Otorhinolaryngol. Head Neck Dis. **132**, 213–215 (2015)
6. IHE PCC Technical Committee, Cross Enterprise Tumor Board Workflow Definition (XTB-WD) Trial Implementation. IHE International (2014)
7. Robertson, T., Li, J., O'Hara, K., Hansen, S.: Collaboration within different settings: a study of co-located and distributed multidisciplinary medical team meetings. Comput. Support. Coop. Work **19**, 483–513 (2010)
8. Look Hong, N.J., Gagliardi, A.R., Bronskill, S.E., Paszat, L.F., Wright, F.C.: Multidisciplinary cancer conferences: exploring obstacles and facilitators to their implementation. J. Oncol. Pract. **6**(2), 61–68 (2010)
9. Tőkés, T., Torgyík, L., Szentmártoni, G., Somlai, K., Tóth, A., Kulka, J., Dank, M.: Primary systemic therapy for breast cancer: does the patient's involvement in decision-making create a new future? Patient Educ. Couns. **98**, 695–703 (2015)
10. Huber, J., Ihrig, A., Winkler, E., Brechtel, A., Friederich, H.C., Herzog, W., Frank, M., Grüllich, C., Hallscheidt, P., Zeier, M., Pahernik, S., Hohenfellner, M.: Interdisciplinary counseling service for renal malignancies: a patient centered approach to raise guideline adherence. Urol. Oncol. Semin. Original Invest. **33**, 23.e1–23.e7 (2015)
11. Choy, E.T., Chiu, A., Butow, P., Young, J., Spillane, A.: A pilot study to evaluate the impact of involving breast cancer patients in the multidisciplinary discussion of their disease and treatment plan. Breast **16**, 178–189 (2007)
12. Munro, A.J., Swartzman, S.: What is a virtual multidisciplinary team (vMDT)? Br. J. Cancer **108**, 2433–2411 (2013)
13. Mangesius, P., Fischer, B., Schabetsberger, T.: An approach for software-driven and standard-based support of cross enterprise tumor boards. In: eHealth - Health Informatics Meets eHealth Proceedings (2015)

14. Sallnäs, E.L., Moll, J., Frykholm, O., Groth, K., Forsslund, J.: Pointing in multi-disciplinary medical meetings. In: International Symposium on Computer Based Medical Systems 24 (2011)

15. Jalil, R., Ahmed, M., Green, J.S.A., Sevdalis, N.: Factors that can make an impact on decision-making and decision implementation in cancer multidisciplinary teams: An interview study of the provider perspective. Int. J. Surg. **11**, 389–394 (2013)

16. Jazieh, A.R.: Tumor boards: beyond the patient care conference. J. Cancer Educ. **26**, 405–408 (2011)

17. Frykholm, O., Nilsson, M., Groth, K., Yngling, A.: Interaction design in a complex context: medical multidisciplinary team meetings. In: NordiCHI (2012)

18. Ashton, M.A.: The multidisciplinary team meeting: how to be an effective participant. Diagn. Histopathol. **14**(10), 519–523 (2008)

19. Stevenson, M.M., Irwin, T., Lowry, T., Ahmed, M.Z., Walden, T.L., Watson, M., Sutton, L.: Development of a virtual multidisciplinary lung cancer tumor board in a community setting. J. Oncol. Pract. **9**(3), e77–e80 (2013)

20. Chekerov, R., Denkert, C., Boehmer, D., Suesse, A., Widing, A., Ruhmland, B., Giese, A., Mustea, A., Lichtenegger, W., Sehouli, J.: Online tumor conference in the clinical management of gynecological cancer: experience from a pilot study in Germany. Int. J. Gynecol. Cancer **18**(1), 1–7 (2008)

21. Parker, B.A., Schwaederlé, M., Scur, M.D., Boles, S.G., Helsten, T., Subramanian, R., Schwab, R.B., Kurzrock, R.: Breast cancer experience of the molecular tumor board at the University of California, San Diego Moores Cancer Center. J. Oncol. Pract. **11**, 442–449 (2015)

22. El Saghir, N.S., Keating, N.L., Carlson, R.W., Khoury, K.E., Fallowfield, L.: Tumor boards: optimizing the structure and improving efficiency of multidisciplinary management of patients with cancer worldwide. Am. Soc. Clin. Oncol., e461-6 (2014)

23. Patkar, V., Acosta, D., Davidson, T., Jones, A., Fox, J., Keshtgar, M.: Cancer multidisciplinary team meetings: evidence, challenges, and the role of clinical decision support technology. Int. J. Breast Cancer (2011)

24. Lamb, B.W., Sevdalis, N., Benn, J., Vincent, C., Green, J.S.A.: Multidisciplinary cancer team meeting structure and treatment decisions: a prospective correlational study. Ann. Surg. Oncol. **20**(3), 715–722 (2013)

25. Lamb, B.W., Brown, K.F., Nagpal, K., Vincent, C., Green, J.S.A., Sevdalis, N.: Quality of care management decisions by multidisciplinary cancer teams: a systematic review. Ann. Surg. Oncol. **18**(8), 2116–2125 (2011)

26. Kane, B.T., Toussaint, P.J., Luz, S.: Shared decision making needs a communication record. In: Proceedings of the 2013 Conference on Computer Supported Cooperative Work, pp. 79–90. ACM (2013)

27. Kane, B., Luz, S.: On record keeping at multidisciplinary team meetings. In: 2011 24th International Symposium on Computer-Based Medical Systems (CBMS), pp. 1–6. IEEE (2011)

28. Avila-Garcia, M.S., Trefethen, A.E., Brady, M., Gleeson, F.: Using interactive and multi-touch technology to support decision making in multidisciplinary team meetings. In: 2010 IEEE 23rd International Symposium on Computer-Based Medical Systems (CBMS), pp. 98–103. IEEE (2010)

29. Kane, B., O'Byrne, K., Luz, S.: Assessing support requirements for multidisciplinary team meetings. In: 2010 IEEE 23rd International Symposium on Computer-Based Medical Systems (CBMS), pp. 110–115. IEEE (2010)

30. Lenz, R., Reichert, M.: IT support for healthcare processes. In: Aalst, W.M.P., Benatallah, B., Casati, F., Curbera, F. (eds.) BPM 2005. LNCS, vol. 3649, pp. 354–363. Springer, Heidelberg (2005)

A Model for Semantic Medical Image Retrieval Applied in a Medical Social Network

Riadh Bouslimi[1,2(✉)], Mouhamed Gaith Ayadi[1,2], and Jalel Akaichi[2]

[1] Higher Institute of Technological Studies, Jendouba, Tunisia
[2] Computer Science Department, BESTMOD Lab,
ISG-University of Tunis, Le Bardo, Tunisia
bouslimi.riadh@gmail.com,
mouhamed.gaith.ayadi@gmail.com, j.akaichi@gmail.com

Abstract. We present in this article a multimodal research model for the retrieval of medical images based on the extracted multimedia information from a radiological collaborative social network. However, opinions shared on a medical image in a medical social network constitute a textual description that requires in most of the time cleaning using a medical thesaurus. In addition, we describe the textual description and medical image in a TF-IDF weight vector using an approach of « bag-of-words ». We use latent semantic analysis to establish relationships between textual and visual terms from the shared opinions on the medical image. Multimodal modeling will search for medical information through multimodal queries. Our model is evaluated on the basis ImageCLEFmed'2015 for which we have the ground-truth. We have carried many experiments with different descriptors and many combinations of modalities. Analysis of the results shows that the model is based on two methods can increase the performance of a research system based on only one modality, either visual or textual.

Keywords: Medical image retrieval · Bag-of-word · Latent semantic analysis · Multimodal fusion · Medical social network

1 Introduction

These Social networks on the theme of health seem to flourish on the Web in recent years. These social networks bring together the health professionals or only doctors and open to industry (pharmaceutical, ..) or to patients. In bottles, these networks their mission is to revolutionize medicine in the real world by accomplishing a connection between doctors, exchange ideas safely without wasting time. Patients can have real-time notifications of doctors and thus have different opinions.

Several medical social networks have been developed in recent years such as: PatientsLikeMe[1], DailyStrength[2], MedPics[3], Carenity[4], etc. which led to an explosion

---

[1] https://www.patientslikeme.com/.
[2] http://www.dailystrength.org/.
[3] https://www.medpics.fr/.
[4] https://www.carenity.com/.

© Springer International Publishing Switzerland 2016
M.E. Renda et al. (Eds.): ITBAM 2016, LNCS 9832, pp. 130–138, 2016.
DOI: 10.1007/978-3-319-43949-5_9

of a number of media collections. The richness and diversity of these collections makes access to useful information increasingly difficult. It has become essential to develop a solution for indexing and research methods appropriate to social networks taking into account the different modalities of contained (text, image, video, etc.).

Content-based medical image retrieval is relative to the context of his capture and interpretation by a radiologist. Current techniques do not allow us to extract images, visual characteristics of low level (color, texture or interest point). The question frequently asked: how best to use the visual characteristics of low level to link automatically to concepts?

This problem is known by the name of "semantic gap" [1]. When both textual and visual modalities are reunited, it will be essential to exploit the two assemblies. Any time, we must take into account that it is easier to automatically associate meaning to a text than to an image.

Many image retrieval systems by the text are realized such as (Google Images, Yahoo!, Flickr...) which are based on information from image annotations. For example, Google indexes web images according the surrounding text (file name, description, link, ...) and Flickr indexes the database images according the keywords that the users assign themselves to images. However, the indexing of text associated with the image is the simplest solution to implement what gives the idea to associate with new images annotations existing in a database [2].

To automatically annotate new images, there are two approaches: the first which is based on a supervised learning and training images are manually classified. Another approach is to automatically discover hidden links between visual and textual elements using unsupervised learning methods [3, 4]. This technique introduces a set of latent variables meant to represent the co-distribution of visual and textual elements.

To annotate a new image, we must first extract the visual description and a function of probabilistic similarity, will return the state that maximizes the probability density of the text annotation and the visual element. Finally, the annotations are sorted by probability values.

Medical information in collaborative medical social networks do not cease to grow, from which comes the necessity to develop medical image retrieval and annotation systems Appropriate self while using both modalities. A natural approach is to use the representation based on "bag of words" for modeling the image. This approach has already proven effective, especially for image medical annotation applications [5, 6]. Indeed, standard collections as TREC, or ImagEVAL and ImageCLEFmed for the evaluation of these systems. We propose a model that combines both visual and textual information in collaborative social networks of health. The relevance of our model is evaluated on a search for medical information. This makes comparing the results obtained with our model to those obtained with a single modality, textual or visual. In this article, we first present our model of representation of information in a social network Medical. Next, we present a semantic medical image annotation model based of multimodal research by applying a fusion by latent analysis. Finally, we show the results of the application on the ImageCLEFMed'2015 dataset.

2 Representation of Information in a Medical Social Network

We present our model of medical social network's representation, which is to describe the text and images with textual terms and visual terms. The two modalities are first processed separately using the approach "bag-of-words" to the visual and textual description. Indeed, they are represented as TF-IDF weight vector characterizing the frequency of each visual or textual words. The vector describing the textual content is cleaned using the UMLS thesaurus. To use the same mode of representation for the two modalities can be combined with a fusion method by latent semantic analysis (LSA), after making multimodal queries to retrieve information. This general method is presented in Fig. 1.

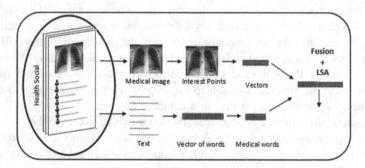

Fig. 1. Representation of content collaborative social health network.

2.1 Representation of Textual Modality

To represent a text report in the form of a weight vector, it is first necessary to define an index of textual or vocabulary terms. For that, we will apply initially a stemming algorithm with Snowball and we delete the black words from all reports. The indexing will be performed by the Lemursoftware[5]. In bottles, the terms selected are then filtered using the thesaurus UMLS. Each report is then represented following the model of Salton [7] is as a weight vector $r_i^T = (w_{i,1}, \ldots, w_{i,j}, \ldots w_{i,|T|})$, where $w_{i,j}$ represents the weight of the term t_j in a report r_i. This weight is calculated as the product of two factors $tf_{i,j}$ and idf_j. The factor tf_{ij} is the frequency of occurrence of the term t_j in the report r_i and the factor idf_j measures the inverse of the frequency of the word in the corpus. Thus, the weight $w_{i,j}$ is even higher than the term t_j, and frequent in the report r_i and rare in the corpus. For the calculation of tf and idf, we are using the formulations defined by Robertson [8]:

---

[5] http://www.lemurproject.org/.

$$tf_{i,j} = \frac{k_1 n_{i,j}}{n_{i,j} + k_2(1 - b + b\frac{|r_i|}{r_{avg}})}$$

where $n_{i,j}$ is the number of occurrences of the term t_j in the report r_i, $|r_i|$ is the report size of r_i and r_{avg} is the average size of all reports in the corpus. k_1, k_2 and b are three constants that take the respective values 1, 1 and 0.5.

$$idf_j = log\frac{|R| - |\{r_i|t_j \in r_i\}| + 0.5}{|\{r_i|t_j \in r_i\}| + 0.5}$$

where $|R|$ is the size of the corpus and $|\{r_i|t_j \in r_i\}|$ is the number of documents in corpus, which the term t_j appears at least once. A textual request q_k can be considered a very short text report, it can also be represented by a weight vector. This vector is noted q_k^T, will be calculated with formulas of Roberson but with b = 0. To calculate the relevance score of a report r_i opposite an request q_k, we apply the formula given by Zhai in [9] and defined as:

$$score_T(q_k, r_i) = \sum_{j=1}^{|T|} r_{i,j}^T q_{k,j}^T$$

2.2   Representation of Medical Image

The representation of the visual modality is carried out in two phases: the creation of a visual vocabulary and the representation of the medical image using thereof. The vocabulary V of the visual modality is obtained using the approach "bag-of-words" [8]. The process consists of three steps: the choice of regions or interest points, the description by calculating a descriptor of points or regions and grouping of descriptors into classes constituting the visual words. We use two different approaches for the first two steps. The first approach uses a regular cutting of the image into n^2 thumbnails. Then a color descriptor with 6 dimensions, denoted Meanstd, is obtained for each thumbnail, by calculating the mean and standard deviation of normalized components $\frac{R}{R+G+B}$, $\frac{G}{R+G+B}$ and $\frac{R+G+B}{3\times 255}$ where R, G and B are the colors components. The second approach uses the characterization of images with regions of interest detected by the MSER [9] and presented by their bounding ellipses (according to the method proposed by [10]). These regions are then described by the descriptor SIFT [11]. For the third step, the grouping of classes is performed by applying the k-means algorithm on the set of descriptors to obtain k clusters descriptors. Each cluster center then represents a visual word. The representation of an image using the vocabulary defined previously for calculating a weight vector r_i^v exactly as for the text modality. To obtain a visual words from the medical image, we first calculate the descriptors on the points or regions of the image, and then is associated, at each descriptor, the word vocabulary, the nearest in the sense of the Euclidean distance.

2.3   Fusion Model

In this section, we show how the two vocabularies are merged using the technique latent semantic.

2.3.1   Latent Semantic Indexing

The Latent Semantic Indexing (LSI) was first introduced in the field of research information and has proven its effectiveness in recent years [12]. This technique involves reducing the indexation matrix in a new space sensible to express more "semantic" dimensions. This reduction is intended to make it appear the "hidden semantics" in the co-occurrence links. This is called latent semantic. This latent semantic allows for example to reduce the effects of synonymy and polysemy. It is also used to index without translation, no dictionary, and parallel corpus, that is to say composed of documents in different languages, but supposed to be translations of each other. Technically, *LSA* method is a matrix transformation operation M of co-occurrence between terms and documents. This is in fact singular value decomposition it[6] of the matrix $M : M_{i,j}$ describes the occurrences of term i in document j. The goal is to compute the matrices U, Σ and V such as:

$$M = U\Sigma V^t$$

- U is the matrix of eigenvectors of MM^t
- V^t is the matrix of eigenvectors of M^tM
- Σ is the diagonal matrix $r \times r$ of the singular values.

This transformation can represent the matrix M as a product of two different sources of information: the matrix U relative to documents and the second matrix ΣV^t relative of terms. Using the k largest Eigenvalues of Σ and by truncating the matrices U and V consequently, we obtain an approximation of k and M:

$$M_k = U_k \sum_k V_k^t$$

where $k < r$ is the dimension of the space latent. This dimension reduction can capture important information and eliminate the least important information that is considered the noise produced by the redundancy of information, such as synonymy or polysemy. It must be noted that the choice of the parameter k is difficult because it must be large enough as not to lose information, and small enough to play the role of redundancy reduction.

2.3.2   Social Medical Report' Research

In this part, for each modality of reports (text and image), it is processed independently. We obtain a textual matrix report-term $M_{r,t}$ and the visual matrix report-term $M_{r,v}$. The fusion of these two methods is first obtained simply by concatenating the columns of

---

[6] Singular Value Decomposition (SVD).

the two matrices $M_{r,t}$ and $M_{r,v}$ in a matrix $M_{r,vt}$ because it is of different coordinates on the same set of documents. This merged matrix is then projected in a latent space to obtain the latent matrix $M_{r,k}$, with k the new reduced size. Therefore, each document is represented by a line latent matrix $M_{d,k}$. For a query containing text and images, we apply the same process with the reports. Then, this vector is projected into the reduced space for a pseudo-vector $q_k = q * \sum_k V_k^t$. Finally, the calculation of the value of relevance of a document to the query (Relevance Status Value or RSV) is calculated according the similarity of the query vector q_k with the lines of the latent matrix by using the function *cosinus*.

3   Experimental Evaluation

3.1   Test Data and Evaluation Criteria

The pertinence of our model is evaluated on the collection provided for the competition ImageCLEFMed'2015 [13]. This collection is composed over 45,000 biomedical research articles of the PubMed Central (R). Each document is composed of an image and a text part. The images are very heterogeneous in size and content. The text part is relatively short with an average of 33 words per document. The goal of the information search task is to return to the 75 queries supplied by ImageCLEFMed'2015 a list of pertinent documents. All requests have a textual part, but many do not have a query image. Order to have a visual part for each query, we use the first two pertinent medical images returned by our system when we use only the textual part. This corresponds to a relevance feedback fact by the system user. The criteria of average accuracy (Mean Average Precision - MAP), which is a classic criteria in information retrieval, is then used to evaluate the pertinence of the results.

3.2   Results and Discussion

To demonstrate the contribution of the use of our model compared to only textual or visual model, we realized experiments using a single modality, textual or visual, then experiments combining two modalities, modality text with visual descriptor, this visual features for both Meanstd and SIFT previously presented. The text vocabulary is consists of approximately 200000 words whereas the two visual vocabularies are constituted of 10000. Table 1 summarizes the MAP values obtained for each experiment. On the one hand, it can be stated that the use of the single visual modality irrespective of the descriptor used leads to poorer results that the use of the only modality text. On the other hand, combine a visual descriptor with the text improves search performance with the only textual descriptor. These overall observations are confirmed by the precision/recall curves presented in Fig. 2. A detailed analysis per request show that, for some, the first results returned by the visual modality is best for text modality. For illustration, Figs. 3, 4 and 5 show the results for the query "Aortic stenosis". We can add, about the performance obtained with a visual modality, that the regular division of the associated image at Meanstd color descriptor is more robust than MSER + SIFT. We explain this behavior by clustering problems. With the color descriptor, we work with 6 characteristic parameters and 4 million thumbnails to

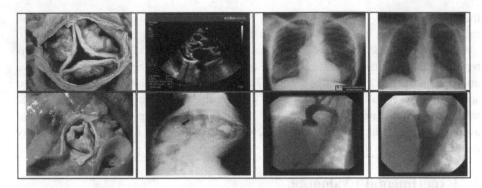

Fig. 2. Results obtained with the textual modality for the query "Aortic stenosis", Figs. 3 and 4 are selected for visual query.

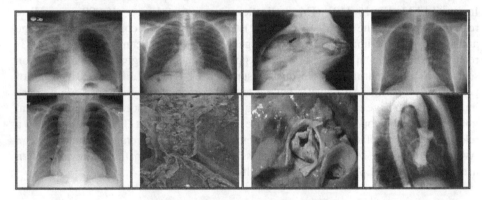

Fig. 3. Results obtained with the visual descriptor Meanstd from the images of the query "Aortic stenosis".

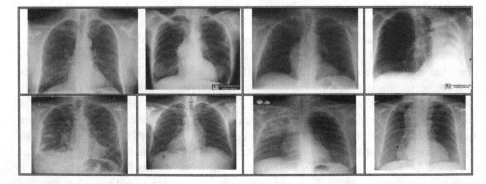

Fig. 4. Results obtained for the modality fusion of the query "Aortic stenosis".

Table 1. Result of average precision obtained for different modalities.

Modality type	MAP
SIFT	0.1287
Meanstd	0.0962
Text	0.2346
Fusion : Text + SIFT	0.3667
Fusion : Text + Meanstd	0.2734

consolidate vocabulary words. With SIFT descriptor, we have 128 features and settings 54 million thumbnails. In the second case, the thumbnails are divided very irregularly in the space of descriptors, because of the use of MSER, the large size and the large amount of data. This situation is very unfavorable for clustering algorithms such as K-means [14]. Also, it has been shown in [15, 16] that the descriptors of the most densely spaces of the parameter space are not necessarily the most informative.

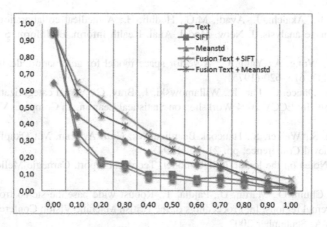

Fig. 5. Precision/Recall curve for different modalities (text only, visual only and fusion text/visual).

4   Conclusion

We have presented in this paper a representing model of multimedia data extracted from radiological social networks. This model is based on a fusion with LSA of the textual and visual information using the "bag-of-words" approach.

The performance of the indexing and search system has been evaluated based on real dataset of ImageCLEFMed'2015 and the obtained results were very promising to use a media model specialized radiology, like the one proposed to retrieve information from a collection of radiological media. Indeed, the fusion of both textual and visual methods allows each time to increase the performance of the system. In this context, Larlus [14] proposes a clustering method that allows uniform quantifications of spaces contrary to K-means which focuses on dense spaces. This method could be used to improve our system when creating the visual vocabulary.

References

1. Priyatharshini, R., Chitrakala, S.: Association based image retrieval: a survey. In: Das, V.V. (ed.) AIM/CCPE 2012. CCIS, vol. 296, pp. 17–26. Springer, Heidelberg (2012)
2. Duan, L., Yuan, B., Wu, C., Li, J., Guo, Q.: Text-image separation and indexing in historic patent document image based on extreme learning machine. In: Cao, J., Mao, K., Cambria, E., Man, Z., Toh, K.-A. (eds.) Proceedings of ELM-2014 Volume 2, PALO, vol. 4, pp. 299–308. Springer, Heidelberg (2014)
3. Clinchant, S., Csurka, G., Ah-Pine, J.: Semantic combination of textual and visual information in multimedia retrieval. In: Proceedings of 1st ACM International Conference Multimedia Retrieval, New York, NY, USA (2011)
4. Wang, S., Pan, P., Lu, Y., Xie, L.: Improving cross-modal and multi-modal retrieval combining content and semantics similarities with probabilistic model. J. Multimedia Tools Appl. **74**(6), 2009–2032 (2013)
5. Bouslimi, R., Akaichi, J.: Automatic medical image annotation on social network of physician collaboration. J. Netw. Model. Anal. Health Inform. Bioinform. **4**(10), 219–228 (2015)
6. Bouslimi, R., Akaichi, J., Ayadi, M.G., Hedhli, H.: A medical collaboration network for medical image analysis. J. Netw. Model. Anal. Health Inform. Bioinform. **5**(10), 145–165 (2016)
7. Salton, G., Wong, A., Yang, C.: A vector space model for automatic indexing. Commun. ACM **18**(11), 613–620 (1975)
8. Csurka, G., Dance, C., Fan, L., Willamowski, J., Bray, C.: Visual categorization with bags of keypoints. In: ECCV 2004 Workshop on Statistical Learning in Computer Vision, pp. 59–74 (2004)
9. Robertson, S., Walker, S., Hancock-Beaulieu, M., Gull, A., Lau, M.: Okapi at trec-3. In: Text REtrieval Conference, pp. 21–30 (1994)
10. Zhai, C.: Notes on the lemur TFIDF model. Technical report, Carnegie Mellon University (2001)
11. Matas, J., Chum, O., Martin, U., Pajdla, T.: Robust wide baseline stereo from maximally stable extremal regions. In: Proceedings of the British Machine Vision Conference, pp. 384–393. BMVA, September 2002
12. Abd Rahman, N., Mabni, Z., Omar, N., Fairuz, H., Hanum, M., Nur Amirah, N., Rahim, T. M.: A parallel latent semantic indexing (LSI) algorithm for malay hadith translated document retrieval. In: First International Conference, SCDS 2015, Putrajaya, Malaysia, pp. 154–163 (2015)
13. de Herrera, A.G.S., Muller, H., Bromuri, S.: Overview of the ImageCLEF 2015 medical classification task. In: Working Notes of CLEF 2015 (Cross Language Evaluation Forum) (2015)
14. Larlus, D., Dorkó, G., Jurie, F.: Création de vocabulaires visuels efficaces pour la catégorisation d'images. In: Reconnaissance des Formes et Intelligence Artificielle (2006)
15. Jurie, F., Triggs, W.: Creating efficient codebooks for visual recognition. In: ICCV 2005 (2005)
16. Vidal-Naquet, M., Ullman, S.: Object recognition with informative features and linear classification. In: ICCV, pp. 281–288 (2003)

Poster Session

A Clinical Case Simulation Tool for Medical Education

Juliano S. Gaspar[1,2,3(✉)], Marcelo R. Santos Jr.[1], and Zilma S.N. Reis[1,2]

[1] CINS – Centro de Informática em Saúde da Faculdade de Medicina da UFMG,
Universidade Federal de Minas Gerais, Belo Horizonte, Brazil
julianogaspar@gmail.com, marrsantosjunior@gmail.com,
zilma.medicina@gmail.com

[2] Departamento de Ginecologia e Obstetrícia da Faculdade de Medicina da Universidade,
Federal de Minas Gerais, Belo Horizonte, Brazil

[3] UNIBH – Centro Universitário de Belo Horizonte, Belo Horizonte, Brazil

Abstract. The human being, even if potentially inclined to learn, needs incentives to do it effectively. In these context, the virtual environments could simulate challenges of clinical practice and, at the same time, consider the personal experiences, allows the student's stimulus and also offer additional theoretical content updated and of excellent quality. The purpose of the project is to develop a Clinical Case Simulation Tool (CCST), it´s supposed to be a supporter to the acquisition of clinical skills for medical education. This is an experimental study of applied technology for health education. The project is multidisciplinary between health sciences, computing and education. The development of an application to store real clinical cases is the starting point of this study. The structure of the proposed clinical case comprises the description of the case, clinical history, complementary tests, questions and further reading. The access to the application is password protected, composed of access profiles with specific characteristics such as teacher, coordinator, student. All clinical cases are linked to a specific college and discipline. The Clinical Cases simulator platform was created for storage of clinical cases and to provide technological support for preparing courses, workshops and support classroom teaching. This may be considered as an innovative approach, given the use of a digital system that enables the storage of clinical data and laboratory tests, as sounds of cardiac auscultation, pulmonary auscultation, images and videos.

Keywords: Computer simulation · Gamification · Competency-Based education · Clinical diagnosis · Medical education

1 Introduction

The human being, even if potentially inclined to learn, needs incentives to do it effectively [1]. Nowadays, a teacher is supposed to be the driver of a teaching-learning process. In the meantime, the student has much more active role to acquire skills and professional competence. These are considered as the major basic challenges of the current education: learning to know, learning to do and learning to be [2].

© Springer International Publishing Switzerland 2016
M.E. Renda et al. (Eds.): ITBAM 2016, LNCS 9832, pp. 141–150, 2016.
DOI: 10.1007/978-3-319-43949-5_10

Every educational system is built to meet the needs of the society that created it. However, with the recent advancement of technology, the educational system today does not match the educational needs of contemporary society [3].

There is a global challenge to adapt the training of health professionals, facing the limitations of traditional models of education [4]. The competency-based education purposes enhancing of new dimensions and skills for care. In order to prepare more capable health professional, the social issues and the relationships between people matter, as well simulation to lower medical mistakes and teaching strategies of problem-solving [5].

The initial training of medical graduate students demands repetition of specific tasks, controlled behavior and adaptation of an individual time of learning. Fundamental clinical skills domain gathers communication, physical examination, clinical reasoning and proposal of diagnostic measures and therapeutic plan. Despite technological advances, the physical examination of an individual by a doctor is a practice considered, besides cost-effective, extremely important for the diagnosis of diseases, establishment of a diagnosis plan, therapeutic and prognostic, in many situations [6]. However, it takes more than good books to acquire the skills necessary to perform a physical examination. In this context, computer simulation environments can contribute by allowing an ethical repetition of tasks and individualized oriented training in support of learning in real scenario [7].

The virtual environments could simulate challenges of clinical practice and, at the same time, consider the personal experiences, allows the student's stimulus and also offer additional theoretical content updated and of excellent quality.

The simulation consists in using specific computational techniques for the purpose to repeat the process or operation in the real world [8]. It is an important tool to compare and evaluate process changes. The effects of changes in processes is evaluated by measuring performances. As the simulation runs multiple iterations, the performance of students are saved for future performance analysis [9].

In medical education, simulation is the replacement of real patient encounters with either standardized patients or technologies that replicate the clinical scenario [10]. The virtual patient simulations are together education games termed gamified training platforms. Kind of technology applied to medical education, those tools are growing and resulting in a good impact for all levels of medical trainees, providers, and educators [11].

The purpose of the project is to develop a learning virtual environment using clinical cases. The Clinical Case Simulation Tool (CCST) gathering variable complexity and attention level settings is supposed to be a supporter to the acquisition of clinical skills among students of health professional.

2   Methods

This is an experimental study of applied technology for health education. The project is multidisciplinary between health sciences, computing and education. The development of an application to store real clinical cases is the starting point of this study. In a first step, undergraduate medical student and professors are collecting real clinical cases. The professor can register the clinical cases with various complexities, whether based on clinical, ambulatory and hospital care. For real clinical cases, patients will be invited to a voluntary participation.

The present study was approved by the Committee for Research Ethics of Federal University of Minas Gerais under the registration number (Brazil Plataform: 10286913.3.0000.5149) all human research principles respected. Using anonymous presentation, the medical history is composed by clinical examination, laboratory tests, whether in reports or sound, video and image file formats. The medical history library is stored in the Faculty of Medicine of UFMG sever. The content is used exclusively for the training of health professionals. Access is password protected and will be under the responsibility of the project coordinator.

The application makes possible to store the specific characteristics of each type of case: text, images, videos and sound files. A short and objective medical report, clinical images such as ultrasound or magnetic resonance and digital electrocardiogram can be stored in the system, as well as cardiac auscultation sounds. In order to give more realistic sensation during the simulation, cardiac sounds are stored with a digital stethoscope are separated by the aortic, pulmonary, mitral and tricuspid.

During the second step of learning preparation, each clinical case will be reviewed. Professors prepare one or more questions related to the case, theoretical or practical questions focused on diagnosis, prognosis or therapeutic process. Each clinical challenge must be properly explained, following of the correct answer and the other about why they are wrong.

The purpose is offering a virtual clinical challenge and stimulate the students to solve virtual patient cases to support the transfer of learning. In a complementary and final step of the learning process, the system can provide specific explanations about the topic and suggested further reading. All the explanations must be based on published, containing citations and references used.

The application must be ergonomic and self-explanatory, so that teachers can simply document the clinical cases of interest to the improvement of health professionals learning.

3   Results

The structure of the proposed clinical case comprises the description of the case, clinical history, complementary tests, questions and further reading (Fig. 1).

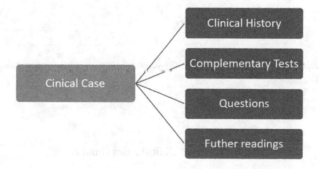

Fig. 1. Structure of clinical cases

In Fig. 2 it can be seen from the CCST database diagram, there are listed the respective tables and fields.

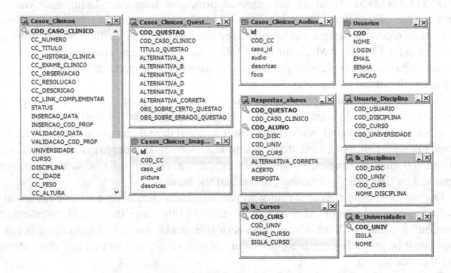

Fig. 2. Data base structure

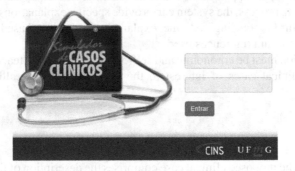

Fig. 3. Access to clinical cases simulator

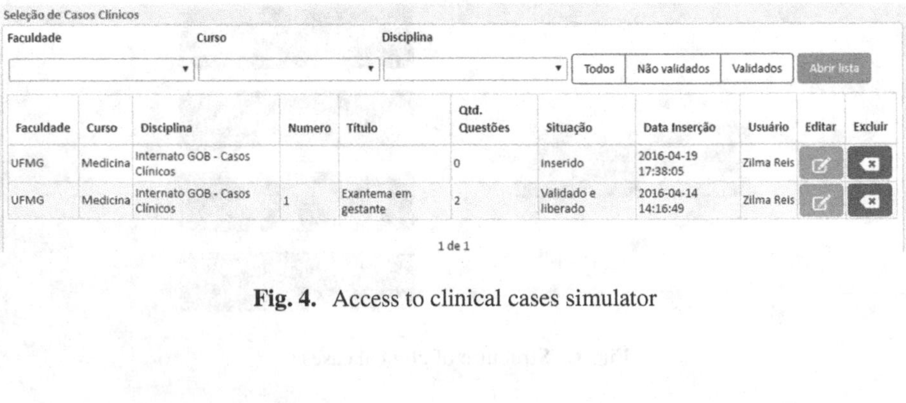

Faculdade	Curso	Disciplina	Numero	Título	Qtd. Questões	Situação	Data Inserção	Usuário	Editar	Excluir
UFMG	Medicina	Internato GOB - Casos Clínicos			0	Inserido	2016-04-19 17:38:05	Zilma Reis		
UFMG	Medicina	Internato GOB - Casos Clínicos	1	Exantema em gestante	2	Validado e liberado	2016-04-14 14:16:49	Zilma Reis		

1 de 1

Fig. 4. Access to clinical cases simulator

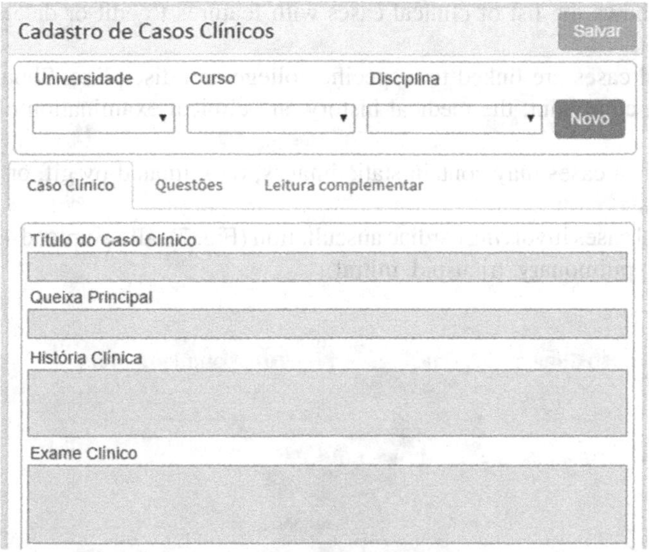

Cadastro de Casos Clínicos
Salvar
Universidade
Curso
Disciplina
Novo
Caso Clínico
Questões
Leitura complementar
Título do Caso Clínico
Queixa Principal
História Clínica
Exame Clínico

Fig. 5. Basic components of a clinical case

The access to the application is password protected (Fig. 3), composed of access profiles with specific characteristics such as teacher, coordinator, student.

Caso Clínico
Questões
Leitura complementar
Imagens

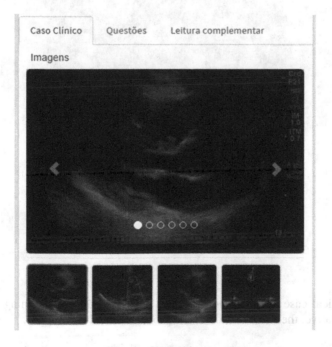

Fig. 6. Clinical images

Figure 4 shows the list of clinical cases with features to edit or delete an already registered case.

All clinical cases are linked to a specific college and discipline. Should contain a title, the main complaint, the medical history and clinical examination of the patient (Fig. 5).

Some clinical cases may contain static images, or animated by gif, or even videos (Fig. 6).

The clinical cases involving cardiac auscultation (Fig. 7), allow upload 4 foci of heart sounds (aortic, pulmonary, tricuspid, mitral).

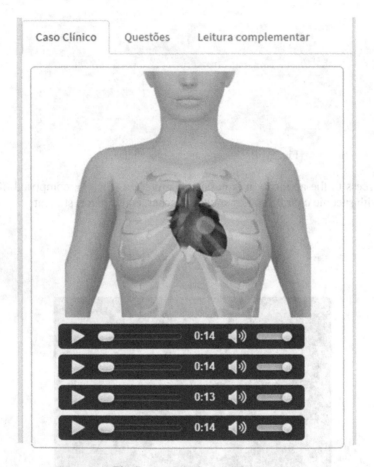

Fig. 7. Auscultation simulator

Each clinical case is composed of one or more questions, containing besides the correct alternative, their explanation of the problems involved (Fig. 8).

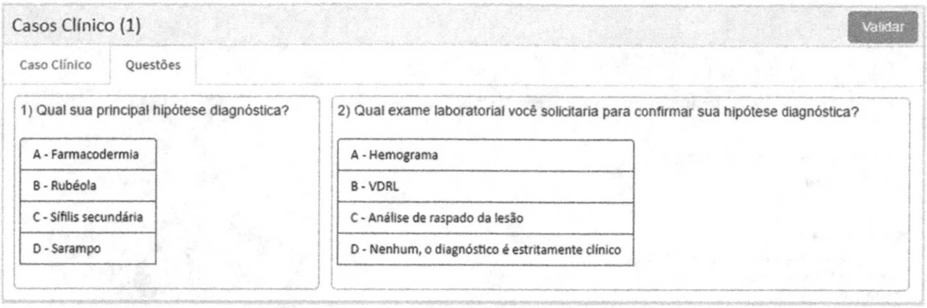

Fig. 8. Clinical case question example

In addition to the clinical case, after the resolution of the issues, the student has access to further reading about the topic, as well as specific explanations about the case in question (Fig. 9).

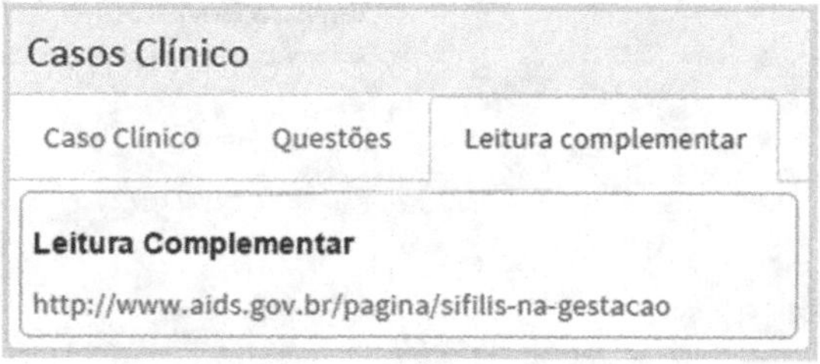

Fig. 9. Supplementary reading

The student, through its specific profile, you can view the list with clinical cases already decided, and the list of new cases. Figure 10 shows the display of a case by the student profile.

In closing, the clinical cases simulator presents a history of resolutions, as well as the score of the general or separate students by discipline (Fig. 11).

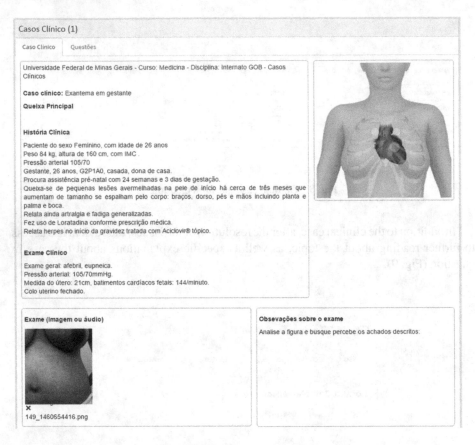

Fig. 10. Supplementary reading

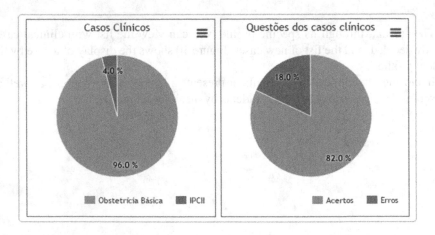

Fig. 11. Student performance

The Clinical Cases Simulator platform can be accessed at the website http://simu-lacao.medicina.ufmg.br/ using the login/password = itbam/itbam2016.

4 Discussion

The Clinical Cases simulator platform was created for storage of clinical cases and to provide technological support for preparing courses, workshops and support classroom teaching, managing the access of students and teachers, the content, and the performance evaluation.

This tool assigns to the teacher the driver role of the teaching-learning process, collecting and developing clinical cases, documenting them in the form of text, sound and images, from real scenarios and register them in a virtual environment for later use as a support tool to education in health care. The teacher have also to prepare the sessions of clinical challenges in a virtual environment, in the form of questions/ answers (quiz) with tracks from reports, images, graphs, videos, and specific theoretical content to support the training of health professionals.

The clinical cases simulators in use at Harvard Medical School, is based on a human patient simulator, as they are high-fidelity humans dolls. The clinical cases are made by teachers and solved by the students, with different approaches, according to your level. The negatives of this educational process is the cost of implementation and operation. In addition to this educational procedure is used, it should have a dedicated local and specific equipment for performing simulations [12].

A similar study was conducted in two US residency programs where medical residents were subjected to a competition using a simulator of clinical cases. The competition relied on general issues of residency and issues three medical specialties. The questions were written by teachers and the correct answers had a wide explanation [13].

This study contributes to the interdisciplinary of the sciences as it brought together a multidisciplinary team of doctors of various specialties and professionals in computer science, adapting to the needs of the development of technical skills in building and maintaining the data collection system, analysis of the functionalities of the system and preparation of content within the various areas of health.

This may be considered as an innovative approach, given the use of a digital system that enables the storage of clinical data and laboratory tests, as sounds of cardiac auscultation, pulmonary auscultation, images, videos. Composing a vast digital library of real cases of extreme importance to education in health, whether in classroom courses or distance, which makes the unprecedented application in the literature.

In the future it is intended to measure the effectiveness of learning or acquiring skills of students by comparing performance tests pre and post use of the Clinical Cases Simulator.

The developed digital library, Clinical Cases Simulator, constitutes the starting point for professional training activities and scientific studies. It is expected to contribute in the future to train more qualified health professionals.

Acknowledgements and Support. CAPES (Coordenação de Aperfeiçoamento de Pessoal de Nível Superior).

FAPEMIG (PPSUS APQ 3486-13) - Fundação de Amparo a Pesquisa de Minas Gerais.

Departamento de Ginecologia e Obstetrícia da Faculdade de Medicina da Universidade Federal de Minas Gerais, Brasil.

References

1. Zeferino, A.M.B., Passeri, S.: Avaliação da aprendizagem do estudante. Cadernos da ABEM **3**, 39–43 (2007)
2. Antunes, C.: Como desenvolver as competências em sala de aula. Editora Vozes Limitada (2012). ISBN 853264208X
3. Rădulescu, C., Bucur, C., Ciolan, L., Petrescu, A.: what does the teacher look like nowadays? CrossCultural Manage. J. **8**, 123–127 (2015)
4. McLeod, P., Steinert, Y., Chalk, C., Cruess, R., Cruess, S., Meterissian, S., Snell, L.: Which pedagogical principles should clinical teachers know? Teachers and education experts disagree Disagreement on important pedagogical principles. Med. Teach. **31**(4), e117–e124 (2009)
5. Cyrino, E.G., Toralles-Pereira, M.L.: Trabalhando com estratégias de ensino-aprendizado por descoberta na área da saúde: a problematização e a aprendizagem baseada em problemas. Cadernos de Saúde Pública **20**, 780–788 (2004)
6. Lederle, F.A., Walker, J.M., Reinke, D.B.: Selective screening for abdominal aortic aneurysms with physical examination and ultrasound. Arch. Int. Med. **148**(8), 1753–1756 (1988)
7. Troncon, L.E.A.: Utilização de pacientes simulados no ensino e na avaliação de habilidades clínicas. Medicina (Ribeirao Preto. Online) **40**(2), 180–191 (2007)
8. Davidsson, P.: Agent based social simulation: A computer science view. J. Artif. Societies Soc. Simul. **5**(1), 305–332 (2002)
9. Benneyan, J.C.: An introduction to using computer simulation in healthcare: patient wait case study. J. Soci. Health Syst. **5**(3), 1–15 (1997)
10. Okuda, Y., Bryson, E.O., DeMaria, S., Jacobson, L., Quinones, J., Shen, B., Levine, A.I.: The utility of simulation in medical education: what is the evidence? Mount Sinai J. Med.: A J. Trans. Personalized Med. **76**(4), 330–343 (2009)
11. McCOY, L., Lewis, J.H., Dalton, D.: Gamification and multimedia for medical education: a landscape review. J. Am. Osteopath. Assoc. **116**(1), 22–34 (2016)
12. Gordon, J.A., Oriol, N.E., Cooper, J.B.: Bringing good teaching cases "to life": a simulator-based medical education service. Acad. Med. **79**(1), 23–27 (2004)
13. Nevin, C.R., Westfall, A.O., Rodriguez, J.M., Dempsey, D.M., Cherrington, A., Roy, B., Willig, J. H.: Gamification as a tool for enhancing graduate medical education. Postgrad. Med. J. Postgradmedj-2013 (2014)

Covariate-Related Structure Extraction from Paired Data

Linfei Zhou[1], Elisabeth Georgii[2], Claudia Plant[3], and Christian Böhm[1(✉)]

[1] Ludwig-Maximilians-Universität München, Munich, Germany
{zhou,boehm}@dbs.ifi.lmu.de
[2] Helmholtz Zentrum München, Neuherberg, Germany
elisabeth.georgii@helmholtz-muenchen.de
[3] University of Vienna, Vienna, Austria
claudia.plant@univie.ac.at

Abstract. In the biological domain, it is more and more common to apply several high-throughput technologies to the same set of samples. We propose a Covariate-Related Structure Extraction approach (CRSE) that explores relationships between different types of high-dimensional molecular data (views) in the context of sample covariate information from the experimental design, for example class membership. Real-world data analysis with an initial pipeline implementation of CRSE shows that the proposed approach successfully captures cross-view structures underlying multiple biologically relevant classification schemes, allowing to predict class labels to unseen examples from either view or across views.

1 Introduction

With the development of modern omics technologies, massive numbers of variables are measured at the same time. For instance, sequencing technologies allow to quantify expression levels for tens of thousands of transcripts. Furthermore, multiple measurement types are frequently co-applied, providing different views on the same biological samples. Multi-view data occur also in other domains, for instance, textual descriptions combined with images of objects. In addition, biological samples often have covariate information attached, which stems from the experimental design. This information can be in form of categorical class labels such as disease group of a patient or in form of numerical variables such as body weight. The approach proposed in this paper takes covariate information into account while analyzing relationships between different data views. The goal is to find such relationships that capture covariate-related structure in the data, for example class separation.

To integrate data from multiple views, a lot of approaches have been proposed, also known as data fusion or multi-block analysis methods (Westerhuis et al. (1998); Smilde et al. (2003); Lanckriet et al. (2004); Tenenhaus and Vinzi (2005); Jiang et al. (2012); Acar et al. (2014)). These methods have been used in various areas (Jamali and Ester (2010); Acar et al. (2012); Lee et al. (2012)).

© Springer International Publishing Switzerland 2016
M.E. Renda et al. (Eds.): ITBAM 2016, LNCS 9832, pp. 151–162, 2016.
DOI: 10.1007/978-3-319-43949-5_11

Multi-block analysis handles multiple blocks of data collected on the same set of samples (Westerhuis et al. (1998); Smilde et al. (2003)). The main objective of multi-block analysis approaches is to find latent variables that explain each block while optimizing the correlation between blocks. The multi-block analysis methods can be classified into three groups: generalized Principal Component Analysis (PCA), Partial Least Squares (PLS) regression and Canonical Correlation Analysis (CCA) methods (Westerhuis et al. (1998); Zhou et al. (2015)). Consensus PCA (CPCA) (Wold et al. (1987)), hierarchical PCA (HPCA)(Wold et al. (1996)), multi-group multi-block PCA (mgmbPCA) (Eslami et al. (2014)) and multiple factor analysis (MFA) (Abdi et al. (2013)) are parts of multi-block family of PCA extensions introducing the concept of using multiple blocks in PCA, which identifies orthogonal directions of largest variance. PLS aims to explain an output dataset based on an input dataset (Geladi and Kowalski (1986)). A PLS approach to multi-block analysis (PLS-MBA) (Tenenhaus and Vinzi (2005)) has been proposed by Tenenhaus and Vinzi. Choi et al. also propose a multi-block PLS (MBPLS) (Choi and Lee (2005)) method as a fault detection and identification approach. CCA (Hotelling (1936); Sweeney et al. (2013); Klami et al. (2013)) is a well-known and widely used method for finding a reciprocal relationship and capturing the common variation between two datasets (Hardoon et al. (2004); Witten et al. (2009)). To handle more than two datasets, many variations of CCA methods have been proposed, such as generalized CCA (gCCA) (Horst (1961); Vía et al. (2007)) and tensor CCA (TCCA) (Luo et al. (2015)).

Related to our goal, there exist previous multi-view approaches that take covariate information into account, such as MultiwayCCA (Huopaniemi et al. (2010)) and Supervised CCA (SCCA) (Witten and Tibshirani (2009); Guo et al. (2013)). MultiwayCCA extends multiway ANOVA to the multi-view case by introducing a Bayesian model, which uses shared variables to describe common variation between both data views. SCCA searches for correlations between the data views that are associated with covariate information. As an extension of mgPCA (Krzanowski (1984)), mgmbPCA seeks common vectors of loadings across classes for each view of variables basing on an iterative algorithm. MultiwayCCA has been designed to deal with datasets where the number of variables is much larger than the number of samples (small-n-large-p problem). However, the dimension reduction step used by MultiwayCCA is very time-consuming for high-dimensional omics datasets. SCCA has a sparsity criterion that allows to deal with small-n-large-p situation, but in practice it is slow on high-dimensional data.

To extend classical CCA for the small-n-large-p problem, there are two principal directions, dimension reduction and penalty approaches, and the second one includes ridge regularization and sparse regularization (Vinod (1976); Saunders et al. (1998)). Regularization methods introduce additional parameters, which leads a new problem, how to set values for these parameters. González uses a standard cross-validation procedure to choose optimal parameters (González et al. 2008)). Huopaniemi et al. employ clustering-type factor analysis as a dimension reduction technique to project the high-dimensional data into latent variables spaces (Huopaniemi et al. (2010)). Like Huopaniemi et al., we apply dimension reduction to tackle the small-n-large-p problem, but as a difference we use the covariate information already in that step. In that way we hope to solve the small size problem efficiently and at the same time effectively capture the covariate-related structure of the data.

The paper is organized as follows. The next section shows a specific pipeline implementation as a simple example of our approach, Covariate-Related Structure Extraction (CRSE). Then the effectiveness of the pipeline is demonstrated by good classification accuracy in several biological applications, which is outperforming other approaches. The final section discusses conclusions and future work.

2 Covariate Information Related Structure Extraction from Multi-view Data

In this section we describe a simple pipeline implementation to illustrate the idea of CRSE. The pipeline consists of two parts: covariate-dependent dimension reduction and canonical correlation analysis. The purpose of the covariate-dependent dimension reduction is on the one hand to solve the small sample size problem for the subsequent multi-view analysis and on the other hand to optimally preserve the class separation or variation of the covariate. Here we use PLS to do this. PLS projects original variables into a latent variable space that maximally explains the covariate information, hence it can reduce high-dimensional data while taking account of covariate information. PLS is applied on each data view separately. On the dimension-reduced data, CCA is applied to extract correlated structure between the data views. Figure 1 illustrates the work-flow of the proposed pipeline. The trained model can be used to project data into a covariate-dependent shared representation of both data views. We evaluate the model by classification of left-out samples in the projected space.

2.1 Partial Least Squares

PLS is a supervised method that simultaneously performs dimension reduction of the input data and regression of the output data (Boulesteix and Strimmer (2007)). In the CRSE pipeline, PLS is applied to each data view separately, using the view as input data and the covariate information as output data. The general underlying PLS model is as follows (Geladi and Kowalski (1986)):

$$X = TP^\top + E$$
$$L = UQ^\top + F^*$$

where X is the $n \times p_x$ matrix representing the data view with original variables and L is the $n \times d$ matrix of covariate data. T and U are $n \times l$ matrices with latent variable representations of the samples, where l is fixed to a small number

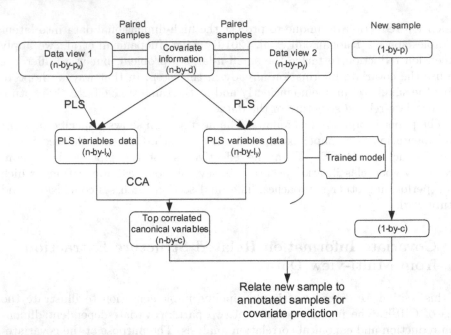

Fig. 1. Diagram of the proposed CRSE pipeline. Each data view is integrated with covariate information by PLS, yielding a low-dimensional representation of the data. The PLS variables are then further processed by CCA to find latent variables capturing shared variation between the views. The standard CCA used here handles two-view data but can be extended to multiple views.

$(l < n < p_x)$. E and F^* are error terms. The overall relation is $L = TBQ^\top + F$, where $\|F\|$ is to be minimized and B relates T and U through regression. By using PLS, we obtain covariate-aware low-dimensional representations of the two views.

2.2   Canonical Correlation Analysis

After applying PLS, the two data views are represented by low-dimensional datasets, which are used in the following step of the pipeline: an $n \times l_x$ matrix T_x and an $n \times l_y$ matrix T_y, where l_x and l_y are the number of latent PLS variables for X and Y, respectively. In the next step of the CRSE pipeline, the relationship between the two low-dimensional data views is analyzed, looking for common variation. Since l_x and l_y do not exceed the sample size (Haenlein and Kaplan (2004)), we can use for that purpose standard CCA without any regularization.

The objective of CCA is to find projections with maximal correlation between the two data views. To obtain the first CCA component, the following objective is solved (Hardoon et al. (2004)):

$$\arg \max_{a^\top T_x^\top T_x a = b^\top T_y^\top T_y b = 1} cor(T_x a, T_y b)$$

where a and b are weight vectors of length l_x and l_y. The subsequence components have the additional constraint that they are uncorrelated to earlier components. This results in two loading matrices A and B of size $l_x \times h$ and $l_y \times h$, where h is the number of paired canonical variables. Thus we get the following canonical variables:

$$C_x = T_x A$$
$$C_y = T_y B$$

Since we are interested in shared variation of the two data views, we focus on the top canonical variables to represent the samples. In practice we choose the canonical variables with a correlation greater than a threshold to focus on what is most common between datasets.

2.3 Data Covariate Prediction

CRSE integrates covariate information with shared components of the data views via two projection steps. Any new sample where we have either one of the data views available can be scaled and projected by the trained PLS plus CCA model. In the projected space, we can apply a classification or regression method to predict the covariate of the new sample, allowing to assess whether the projected representation captures relevant information from the samples. Remarkably, when predicting the covariate label of a new sample in a data view, the common space allows to not only use the labeled samples of this data view but also all the labels of the other data view, including in particular non-paired labeled samples.

3 Experiments and Results

To demonstrate the effectiveness and efficiency of the CRSE pipeline, we make experiments on real-world datasets from the model plant *Arabidopsis thaliana*. Each dataset consists of two data views, a metabolomic data view and a gene expression data view. As covariate information, the biological samples are annotated by two different classification schemes, genotype and environmental condition. Genes modified in the genotypes, allowing a trivial genotype separation, were excluded from the expression data. Table 1 summarizes the key properties of both datasets.

The datasets are preprocessed using the R packages limma (Smyth (2004); Ritchie et al. (2015)), FTICRMS (Barkauskas (2012)) and nlme (Pinheiro and Bates (2000)). The CRSE pipeline was implemented based on built-in functions in R. All the experiments are executed on a regular workstation PC with 3.4 GHz dual core CPU equipped with 32 GB RAM.

The canonical variables are the projections of the original data in a new space that represents the maximum correlation structure between views and preserves covariate variation. Assuming that canonical variables keep the principal information and the basic structure of the original data, the classification result of

Table 1. Paired datasets

	Dataset 1	Dataset 2
Nr. of paired observations	57	23
Nr. of variables in metabolomic view	1454	203
Nr. of variables in expression view	24603	24603
Nr. of classes in genotype covariate	3	2
Nr. of classes in condition covariate	6	2

the sample objects in the new space should have a similar accuracy to that of original data. Since irrelevant information and noise might be cleaned out, the accuracy in the new space could be even better. So we use a classification-based method to evaluate the effectiveness of the canonical variables and check whether the detected cross-view relationships are meaningful.

For the analysis we consider paired samples between expression data and metabolome data. After applying scaling to make each variable in the training dataset to have a mean of zero and a standard deviation of one, we reuse the scaling parameters on test samples. Since we have categorical covariate information, we employ dummy coding, which uses only ones and zeros to convert all label information (Wendorf (2004); Boulesteix and Strimmer (2007)). Assuming there are k groups, $k - 1$ dummy coding variables are needed to represent each group. The evaluation is performed by leave-one-out analysis, i.e., n-fold cross-validation where n equals the number of samples. In each round, all data except one sample are used to train a model, and then we apply the model on the left-out sample to get the canonical variables. A k-Nearest Neighbor (k-NN)) classifier (Duda et al. (1973)) is applied on the canonical variable representation to predict a covariate label for the left-out sample. The accuracies of classification on original data and canonical variables will be compared to evaluate the CRSE model. The number of PLS latent variables in CRSE are chosen by nested cross-validation on the training data (i.e., not touching the test data), using the classification accuracy.

We compare the results of CRSE with that of MultiwayCCA and SCCA. Considering that the factor analysis of MultiwayCCA is too expensive and it cannot finish in a reasonable time for our data (it takes hours to finish pilot experiments on an example dataset with only 1000 variables in the expression view), we use k-means as an alternative to reduce the variable number. In analogy to the MultiwayCCA dimension reduction, we cluster all the variables of the original data into k clusters, where $k < n$, and then use the cluster centers as low-dimensional latent variables for the following procedure. Nevertheless, MultiwayCCA still costs much more time than CRSE. Since SCCA is very slow on the full high-dimensional data, too, we use the clustered data also as input for SCCA, and for comparison purposes also with CCA. In order to prove that the clustering step has not lost much information of the original data with respect to covariate structure, we also evaluate the classification accuracy of the clustered

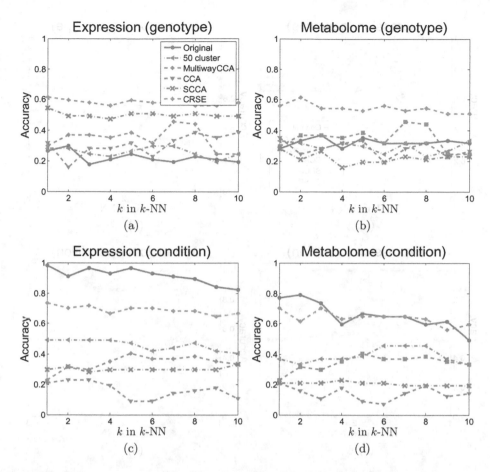

Fig. 2. Classification-based evaluation results on dataset 1 for two types of covariates: genotype and condition. Each sub-figure contains classification accuracies of original data, 50 cluster data, MultiwayCCA output, CCA output, SCCA output and canonical variable output of CRSE. (a) and (b) are the classification results for the three different genotypes, and (a) takes test samples from the expression view while (b) takes that from the metabolomic view. Classification results for the six different conditions are shown in (c) and (d), which indicate expression view and metabolomic view, respectively.

data. Since MultiwayCCA yields by default the first shared component, we apply the k-NN algorithm on the first pair of canonical variables for all the approaches.

The classification results of dataset 1 is shown in Fig. 2. All the classification accuracies have a relatively stable trend with the increase of k in k-NN. In Fig. 2(a) and (b), CRSE has the best performance on classification accuracy, even better than the original data. MultiwayCCA outperforms the clustered data for both views, which achieve similar classification accuracies with the original data. SCCA has the second highest accuracy in Fig. 2(a) but the worst in (b). Combining expression and metabolome with different conditions (Fig. 2(c) and

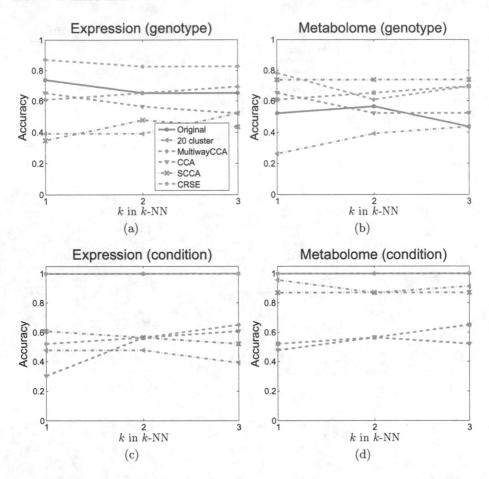

Fig. 3. Classification-based evaluation results on dataset 2 for two types of covariates: genotype and condition. Each sub-figure contains classification accuracies of original data, 20 cluster data, MultiwayCCA output, CCA output, SCCA output and canonical variable output of CRSE. (a) and (b) are the classification results for the two different genotypes, and (a) takes test samples from the expression view while (b) takes that from the metabolomic view. Classification results for the two different conditions are shown in (c) and (d), which indicate expression view and metabolomic view, respectively.

(d)), CRSE achieves no better classification accuracies than the original data, but it is still the best performing method among comparisons. MultiwayCCA, CCA and SCCA show lower accuracies than that of clustered data. The reason for the low performance of all methods compared to the original data might be the complex six-group structure of the condition covariate, which cannot be captured by a single component. This is further analyzed below.

Figure 3 shows the classification results on dataset 2, and the performance of all methods are similar with that of Fig. 2. CRSE achieves the highest classi-

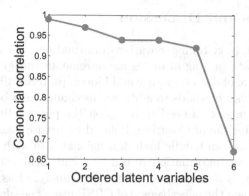

Fig. 4. Correlations of canonical variables in CRSE.

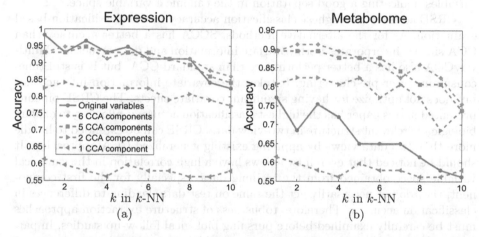

Fig. 5. Classification results using different numbers of latent variables. In this figure, the classification accuracies of experiment on dataset 1 under the six different conditions are shown. (a) and (b) show the classification results on the two data views, respectively.

fication accuracy in most cases, and its accuracy reaches the optimum for both expression data and metabolome data with the condition covariate information.

Since more than one canonical variables have very high correlation values, which are greater than 0.9 (see Fig. 4), we use more canonical variables to do the classification-based evaluation, and the comparisons are shown in Fig. 5. One, two, five and six canonical variables in CRSE are chosen to apply the k-NN algorithm, respectively. The classification accuracy is getting better and better with the increased number of latent variables. When five and six components are used, the accuracy reaches the accuracy of the original data in Fig. 5(a) and it achieves a stable but higher level than that of the original data in Fig. 5(b). This confirms that with a larger number of components, the relevant structure of the six-group condition covariate is successfully captured.

4   Conclusions and Discussion

Multi-view data analysis taking covariate information into account plays an important role in data mining of omics measurements and gives us a potential way to get fully aware of data structure and hidden patterns. Since there is still a lack of efficient analysis methods to address this challenge, a better understanding of the available models is needed to exploit the potentialities. In this paper we have built the pipeline of Covariate-Related Structure Extraction on paired datasets. In CRSE, we can handle high-dimensional data with small sample size and integrate covariate information into a new canonical variable space. Real datasets of *Arabidopsis thaliana* plants have been analyzed as a demonstration, and we have also shown the effectiveness of CRSE using classification-based evaluation of the extracted relationships between metabolome and gene expression variables, indicating a good separation in the canonical variable space.

CRSE achieves the highest classification accuracy in the classification-based evaluation. As for the alternative methods, SCCA has a better accuracy than CCA since it incorporates both covariate information and data structure. Multi-wayCCA also has a better performance than standard CCA, but it is still time-consuming after pre-clustering. It takes the covariate information into account but does not optimize for having significant covariate effects. The CRSE pipeline presented in this paper has the highest classification accuracy, allowing to explore biologically relevant structure in two-view data. CRSE can be extended to handle more than two data views by applying existing generalized CCA approaches. It should be noticed that even if two views have a high correlation in the canonical variable space obtained from the training data, especially for the first component, they do not necessarily act the same on test data, leading to differences in classification accuracy. Therefore, robustness of structure extraction approaches must be carefully examined before pursuing biological follow-up studies. In particular for complex multi-group covariates, a higher classification accuracy can be obtained if more components are chosen.

An advantage of the CRSE approach is that the dimension reduction step of the pipeline can exploit also samples available only for one data view (i.e., non-paired samples). Furthermore, it is straightforward to apply the presented approach with sparse PLS and CCA variants to improve the interpretability of components, or to replace PLS by other dimension reduction methods. From a biological perspective, it will be useful to include known gene-metabolite connections from metabolic pathways into the multi-view model and infer additional relationships for covariate structure that cannot be explained by current knowledge.

Acknowledgement. We thank Ming Jin, Jin Zhao, Basem Kanawati, Philippe Schmitt-Kopplin, Andreas Albert, J. Barbro Winkler, and Anton R. Schäffner for kindly providing the datasets used in this study.

References

Abdi, H., Williams, L.J., Valentin, D.: Multiple factor analysis: principal component analysis for multitable and multiblock data sets. Wiley Interdisc. Rev. Comput. Stat. 5(2), 149–179 (2013)

Acar, E., Gurdeniz, G., Rasmussen, M., Rago, D., Dragsted, L.O., Bro, R.: Coupled matrix factorization with sparse factors to identify potential biomarkers in metabolomics. In: IEEE 12th International Conference on Data Mining Workshops, pp. 1–8 (2012)

Acar, E., Papalexakis, E.E., Rasmussen, M.A., Lawaetz, A.J., Nilsson, M., Bro, R.: Structure-revealing data fusion. BMC Bioinf. 15(1), 239 (2014)

Barkauskas, D.: FTICRMS: Programs for Analyzing Fourier Transform-Ion Cyclotron Resonance Mass Spectrometry Data. R package version 8 (2012)

Boulesteix, A.-L., Strimmer, K.: Partial least squares: a versatile tool for the analysis of high-dimensional genomic data. Briefings Bioinform. 8(1), 32–44 (2007)

Choi, S.W., Lee, I.-B.: Multiblock PLS-based localized process diagnosis. J. Process Control 15(3), 295–306 (2005)

Duda, R.O., Hart, P.E., et al.: Pattern Classification and Scene Analysis, vol. 3. Wiley, New York (1973)

Eslami, A., Qannari, E., Kohler, A., Bougeard, S.: Multivariate analysis of multiblock and multigroup data. Chemometr. Intell. Lab. Syst. 133, 63–69 (2014)

Geladi, P., Kowalski, B.R.: Partial least-squares regression: a tutorial. Anal. Chim. Acta 185, 1–17 (1986)

González, I., Déjean, S., Martin, P.G., Baccini, A., et al.: CCA: an R package to extend canonical correlation analysis. J. Stat. Softw. 23(12), 1–14 (2008)

Guo, S., Ruan, Q., Wang, Z., Liu, S.: Facial expression recognition using spectral supervised canonical correlation analysis. J. Comput. Inf. Sci. Eng. 29(5), 907–924 (2013)

Haenlein, M., Kaplan, A.M.: A beginner's guide to partial least squares analysis. Underst. Stat. 3(4), 283–297 (2004)

Hardoon, D.R., Szedmak, S., Shawe-Taylor, J.: Canonical correlation analysis: an overview with application to learning methods. Neural Comput. 16(12), 2639–2664 (2004)

Horst, P.: Generalized canonical correlations and their applications to experimental data. J. Clin. Psychol. 17(4), 331–347 (1961)

Hotelling, H.: Relations between two sets of variates. Biometrika 28, 321–377 (1936)

Huopaniemi, I., Suvitaival, T., Nikkilä, J., Orešič, M., Kaski, S.: Multivariate multi-way analysis of multi-source data. Bioinformatics 26(12), i391–i398 (2010)

Jamali, M., Ester, M.: A matrix factorization technique with trust propagation for recommendation in social networks. In: Proceedings of the 4th ACM Conference on Recommender Systems, pp. 135–142. ACM (2010)

Jiang, M., Cui, P., Liu, R., Yang, Q., Wang, F., Zhu, W., Yang, S.: Social contextual recommendation. In: Proceedings of the 21st ACM International Conference on Information and Knowledge Management, pp. 45–54. ACM (2012)

Klami, A., Virtanen, S., Kaski, S.: Bayesian canonical correlation analysis. J. Mach. Learn. Res. 14(1), 965–1003 (2013)

Krzanowski, W.: Principal component analysis in the presence of group structure. Appl. Stat. 33, 164–168 (1984)

Lanckriet, G.R., De Bie, T., Cristianini, N., Jordan, M.I., Noble, W.S.: A statistical framework for genomic data fusion. Bioinformatics 20(16), 2626–2635 (2004)

Lee, C.M., Mudaliar, M.A., Haggart, D., Wolf, C.R., Miele, G., Vass, J.K., Higham, D.J., Crowther, D.: Simultaneous non-negative matrix factorization for multiple large scale gene expression datasets in toxicology. PLoS ONE **7**(12), e48238 (2012)

Luo, Y., Tao, D., Ramamohanarao, K., Xu, C., Wen, Y.: Tensor canonical correlation analysis for multi-view dimension reduction. IEEE Trans. Knowl. Data Eng. **27**(11), 3111–3124 (2015)

Pinheiro, J.C., Bates, D.M.: Basic concepts and examples. Mixed-effects Models in S and S-Plus, pp. 3–56 (2000)

Ritchie, M.E., Phipson, B., Wu, D., Hu, Y., Law, C.W., Shi, W., Smyth, G.K.: Limma powers differential expression analyses for RNA-sequencing and microarray studies. Nucleic Acids Res. **43**(7), e47 (2015)

Saunders, C., Gammerman, A., Vovk, V.: Ridge regression learning algorithm in dual variables. In: Proceedings of the 15th International Conference on Machine Learning, pp. 515–521. Morgan Kaufmann (1998)

Smilde, A.K., Westerhuis, J.A., de Jong, S.: A framework for sequential multiblock component methods. J. Chemom. **17**(6), 323–337 (2003)

Smyth, G.K.: Linear models and empirical bayes methods for assessing differential expression in microarray experiments. Stat. Appl. Genet. Mol. Biol. **3**(1), 1–25 (2004). doi:10.2202/1544-6115.1027. ISSN (Online) 1544-6115

Sweeney, K.T., McLoone, S.F., Ward, T.E.: The use of ensemble empirical mode decomposition with canonical correlation analysis as a novel artifact removal technique. IEEE Trans. Biomed. Eng. **60**(1), 97–105 (2013)

Tenenhaus, M., Vinzi, V.E.: PLS regression, PLS path modeling and generalized procrustean analysis: a combined approach for multiblock analysis. J. Chemom. **19**(3), 145–153 (2005)

Vía, J., Santamaría, I., Pérez, J.: A learning algorithm for adaptive canonical correlation analysis of several data sets. Neural Netw. **20**(1), 139–152 (2007)

Vinod, H.D.: Canonical ridge and econometrics of joint production. J. Econometrics **4**(2), 147–166 (1976)

Wendorf, C.A.: Primer on multiple regression coding: common forms and the additional case of repeated contrasts. Underst. Stat. **3**(1), 47–57 (2004)

Westerhuis, J.A., Kourti, T., MacGregor, J.F.: Analysis of multiblock and hierarchical PCA and PLS models. J. Chemom. **12**(5), 301–321 (1998)

Witten, D.M., Tibshirani, R., Hastie, T.: A penalized matrix decomposition, with applications to sparse principal components and canonical correlation analysis. Biostatistics **10**(3), 515–534 (2009)

Witten, D.M., Tibshirani, R.J.: Extensions of sparse canonical correlation analysis with applications to genomic data. Stat. Appl. Genet. Mol. Biol. **8**(1), 1–27 (2009)

Wold, S., Hellberg, S., Lundstedt, T., Sjöström, M.: PLS modeling with latent variables in two or more dimensions. Partial Least Squares Model Building: Theory and Application (1987)

Wold, S., Kettaneh, N., Tjessem, K.: Hierarchical multiblock PLS and PC models for easier model interpretation and as an alternative to variable selection. J. Chemom. **10**(5–6), 463–482 (1996)

Zhou, G., Cichocki, A., Zhang, Y., Mandic, D.P.: Group component analysis for multiblock data: common and individual feature extraction. IEEE Trans. Neural Netw. Learn. Syst. **PP**(99), 1–14 (2015). doi:10.1109/TNNLS.2015.2487364

Semantic Annotation of Medical Documents in CDA Context

Diego Monti(✉) and Maurizio Morisio

Dipartimento di Automatica e Informatica, Politecnico di Torino, Turin, Italy
diegomichele.monti@studenti.polito.it

Abstract. The goal of this work is to recover semantic and structural information from medical documents in electronic format.

Despite the progressive diffusion of Electronic Health Record systems, a lot of medical information, also for legacy reasons, is available to patients and physicians in image-only or textual format. The difficulties of obtaining such information when needed result in high costs for health providers.

In this work we develop the concept of a system designed to convert legacy medical documents into a standard and interoperable format compliant with the Clinical Document Architecture model by the means of semantic annotation.

1 Motivation

In the healthcare domain different kinds of medical documents are produced by physicians in narrative form, relying on templates based on the scope of the document (e.g., progress note, discharge summary). Such templates are slightly different according to the healthcare provider.

Even if an Electronic Health Record (EHR) system is in place, clinical documents usually consist of free text blocks with no semantic encoding. When these documents are exported to a standard electronic format, like a PDF file, in order to send them to patients or other providers, all semantic and structural information is lost.

The difficulties caused by the exchange of medical information among different entities result in high costs for healthcare providers and time consuming activities for patients, that need to act as couriers and perform several times the same medical tests [11]. 25 billion dollars per year are spent in the USA because of unnecessary exams [4].

According to the European and Italian law all medical documents should be published in an interoperable format, but in practice this is not so common. Legacy documents, that contain the medical history of the previous 10–20 years, were initially produced in an electronic format, but nowadays are typically available only in a printed version.

These reasons justify the need of reconstructing the structure of medical documents and performing a semantic annotation on them.

© Springer International Publishing Switzerland 2016
M.E. Renda et al. (Eds.): ITBAM 2016, LNCS 9832, pp. 163–172, 2016.
DOI: 10.1007/978-3-319-43949-5_12

2   Related Work

2.1   Extraction of Semantic Information

Recognizing the structure of paragraphs in image only documents is a well-known problem and the proposed solutions are based on the analysis of the font size and the text placement [2,12]. Also the task of discovering the layout of a textual PDF file has been considered in literature [7]. These studies do not take into account the peculiar characteristics of medical documents, in which the division in sections may be more difficult to identify. On the other side medical documents contain semantic information that can be useful to solve this problem.

The extraction of information from textual medical documents is a very active field of research. Different studies take into account the task of processing non standardized medical data considering text mining and statistical methods in order to identify medical concepts [9], also exploiting a human feedback [20].

A popular approach deals with using Natural Language Processing (NLP) techniques to extract codes, mapping entities present in the text to medical ontologies [13]. cTAKES is an Apache project that aims to extract information from medical documents using the UMLS meta-thesaurus [16]. MetaMap is a program designed to discover UMLS concepts in biomedical text with indexing purposes [1]. MedEx is a medical information extraction system based on the Unstructured Information Management Architecture (UIMA) framework [19].

The limitations of these projects is that they work only with English documents, while we deal with documents in Italian, and that they are not designed with the objective of storing the result of the analysis in a standard medical format.

2.2   Clinical Document Architecture

The Clinical Document Architecture (CDA) is an XML based exchange model for clinical documents proposed by Health Level Seven International, a standards developing organization [8]. The purpose of this model is to enable the sharing of structured medical data among EHR systems of different providers. Three levels of semantic interoperability are defined by the standard.

A CDA document is composed of a header and a body. A level 1 compliant CDA document consists of a free text or an image with some metadata: the body is unstructured and the information about the author and the patience is located in the header. Level 2 compliance means that the body is structured in sections. A CDA document is level 3 compliant if the clinical markup of the body is semantically coded using healthcare code sets [17].

3   Approach

The purpose of this work is to develop a system able to perform an automatic conversion from the PDF version of a clinical document to a CDA level 3 compliant XML file. This operation is organized in three sequential steps (Fig. 1).

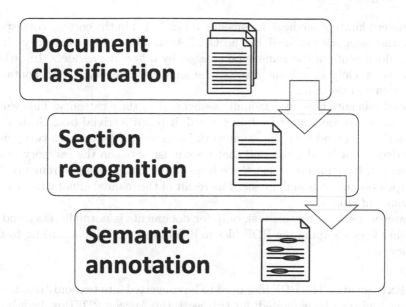

Fig. 1. Graphical representation of the process

3.1 Document Classification

The document is classified in one of the document level categories specified in the Consolidated CDA (C-CDA) standard, a library of CDA templates [10]. Examples of such templates are:

- *Consultation Note*: contains an opinion or advice from another clinician or is used to summarize an Emergency Room or Urgent Care encounter.
- *Discharge Summary*: contains information related to the admission of a patient to a hospital and to the care needed following the discharge.
- *Operative Note*: contains the report of a surgical or other high-risk procedure.
- *Progress Note*: contains a patient's clinical status during a hospitalization, outpatient visit or other healthcare encounter.
- *Transfer Summary*: contains critical information that needs to be exchanged between different providers when a patient moves between health care settings.

This operation is performed by the means of different text mining techniques, in order to select the most effective one. The classifiers considered are: N-gram-based, Naïve Bayes and Bernoulli.

Corpus Annotation. The corpus of input documents is a set of PDF files containing textual information. The following techniques can also be applied to image only documents, but first they must be converted to textual documents employing optical character recognition algorithms.

Different kinds of medical documents are included in the corpus, corresponding to the categories defined in the C-CDA standard. Each category should contain documents of the same type created by different providers. In order to obtain statistically significant results, the smallest category should contain at least a hundred documents.

The documents must be manually associated to their category. This work is known as *corpus annotation* and, in general, it is not a trivial task. State of the art practices demand that the same work is performed by at least two persons knowledgeable of the domain and that they must agree on the category of each document. If for some document there is no agreement, it must be removed from the corpus because it is ambiguous. The result of the manual annotation is called *gold standard* [15].

However, performing this task only for documents is normally easy and can be as simple as grouping the PDF files in different directories according to their category.

Text Extraction. The PDF files need to be converted into text only documents. Different tools can be exploited: for this work the Apache PDFBox Java library has been used.

Medical documents typically contain a header and a footer on each page that do not carry useful information, but they break the flow of the extracted text with, for example, the page number.

PDFBox gives the possibility to extract text only from a specific part of the page by defining a rectangle over the part that will be processed. Considering that a regular A4 page measures 595×842 points, it is generally advisable to avoid analyzing the first and the last 40 points in height. This, of course, needs to be checked for each document.

Classifier Training. LingPipe is a Java library *for processing text using computational linguistics*. Among other features, it is a robust framework to perform automatic text classification [5].

A classifier is trained specifying the possible categories and providing examples for each category. After the training it is possible to compile the classifier in order to produce a more efficient version. The compiled version cannot be trained anymore, but of course can be used to classify new documents. A compiled classifier can be serialized in a file and loaded in a future execution of the program.

Different classifiers are available in LingPipe. The simplest classifier is based on the sequences of characters available in the text: the user needs only to specify the number of characters to analyze together. This kind of classifier is successfully used to perform language recognition, but it is typically too simple to classify documents [6].

Classifiers more suitable for the proposed work are based on the concept of token, so the input text needs to be transformed into a set of tokens. The simplest approach is splitting the text using as separators whitespaces and punctuation characters, transforming all tokens in lowercase characters and discarding short tokens (for example less than four characters) because they probably represent common words. More complex techniques involve statistical tokenizers that are specific for a particular language.

The most famous classifier is the Naïve Bayes classifier. The *naïve* assumption is that each token is independent from other tokens. Despite the name, this classifier usually obtains good results. A variant of this classifier is the Bernoulli model, where for each document only the presence or the absence of a token is considered, not the number of times it appears.

Classifier Validation. The results of different classifiers need to be checked and compared using a technique known as cross-validation.

The training corpus is initially permuted in a random way in order to remove local dependencies. The corpus is then partitioned in a fixed number of folds, typically ten. Each classifier is trained on nine folds and tested on the tenth fold. This operation is repeated ten times, continuously changing the training folds and the test fold. Each time a confusion matrix is computed and for each classifier a mean confusion matrix is created.

The results of different classifiers are finally compared, considering also possible variations in the initialization parameters of each classifier and how the tokens generation is performed.

3.2   Section Recognition

The document is subsequently split into paragraphs analyzing the typographical features available in the PDF file. From the first paragraph the demographic information about the patient, the physician and the provider are extracted using deterministic rules and simple knowledge bases. The following paragraphs are mapped to the section level categories of the C-CDA standard, exploiting the results of the previous classification, statistical classifiers trained for this task and deterministic rules.

Paragraph Extraction. The task of identifying paragraphs in a PDF file is particularly challenging because typically no information about the organization of the document is provided [7].

A PDF file is similar to a vector image: the position of each element is expressed using absolute coordinates. PDFBox supposes that two elements with the same y coordinate are part of the same line. When a line ends it is, in general, impossible to say if the sentence ended or no more space is available on that line.

A PDF file may contain information about its logical structure: this is extremely important for accessibility tools used by blind people. Unfortunately, such tags are typically not present in medical documents if they are exported from a EHR system.

A possible solution to the problem consists in identifying as paragraph title the lines that are written with a different character, but in rare cases it is also possible that the titles are written with the same character of the text.

Another solution is to consider the frequencies of distinct lines for the same category of documents: lines that appear exactly the number of available documents are likely to be paragraph titles.

CDA Header. The first paragraph of a medical document corresponds to the header of the CDA model. The typical problem is that not all the information required by the standard is available in the analyzed file. If possible, a default value should be assigned in this case.

An example of information that is not present in the analyzed file is the unique identifier of the document. The CDA specification requires that an identifier is assigned to the provider using the Object IDentifier (OID) standard and that the provider assigns a unique code to each document that produces [14].

The CDA header also contains the kind of medical document, the time of redaction, the confidentiality level and the language code. The kind of medical document is the result of the previous classification and must be expressed as a code of the LOINC code system. The time of redaction is typically the last date present in the document. For the confidentiality level and the language code it is necessary to use default values.

Finally, the CDA header contains information about the patient, the physician and the provider. For each of them it is necessary to specify a unique code, a name, an address and a telephone number. For the patient also the birthdate, the birthplace and the gender are required. A lot of other information can be optionally added and other fields may be required according to the locale.

Typically, all the information regarding the patient is present, while the data about the provider is buried in an image part of the template and the physician is identified only by his name. The most effective way of discovering the fields of interest is to use regular expressions or a similar formal language designed considering the header of medical documents from different providers.

The usage of simple knowledge bases should be considered to retrieve information that is present in the document only in an implicit way. For example, a list of all possible cities can be exploited to map a birthplace to the corresponding postal code.

Paragraph Classification. The C-CDA standard specifies, for each type of medical document, a list of possible paragraphs. The classification of the remaining paragraphs is therefore linked to the result of the previous classification.

Two different techniques can be adopted: statistical classifiers similar to the ones used to perform the first step of the analysis or a deterministic technique based on the presence of specific keywords in the title or the text of the paragraph.

The usage of a statistical classifier is meaningful only if it was possible to automatically divide the document in sections. A different paragraph classifier is needed for each category of medical documents: if the titles are similar it is advisable to prefer the deterministic method.

In rare situations it is possible that a paragraph has not a corresponding section in the CDA template: this typically means that two or more paragraphs of the analyzed document need to be merged together.

CDA Body. The body of a CDA document consists of a list of sections. Each section must be identified by the relative code in the LOINC code system, discovered thanks to the previous classification. At least the text of the corresponding paragraph in the processed document must be added to the section [10].

It is advisable that the text is correctly divided into sentences. In order to achieve this, the most general solution is to remove any new line character and then to apply a sentence detector. LingPipe only implements deterministic sentence detectors, but also statistically based ones are available in other libraries. In any case better results are obtained if the used technique is dependent on the language.

This solution will work only if it is applied to blocks of text: it is also necessary to consider the possibility that the original paragraph consists of a list of short statements. In this case the new lines should not be removed and every statement should be considered a sentence.

3.3   Semantic Annotation

The text of each paragraph is finally analyzed using NLP techniques in order to identify key concepts (e.g., diseases, procedures, drugs) and to map them to the most appropriate medical ontology, such as LOINC or SNOMED CT. The entities that can be recognized in each block depend on the guessed category of the section. All the extracted information is serialized in a XML file following the CDA specifications.

Code Systems. The first code systems in the medical domain were created for financial reasons, to have procedures justified by a diagnosis in the claims for the health insurance. Nowadays they are also essential to describe laboratory analysis results and prescriptions of drugs without ambiguity [18].

An ontology is a code system where the relationships among different entities are represented in a graph structure. Each entity is at least characterized by a code, a normalized name, a list of possible synonyms and the code of the entity of which it is a generalization or a specialization.

Different medical ontologies are available: the most famous ones are the International Classification of Diseases (ICD), SNOMED CT, LOINC and RxNorm. ICD is a list of diseases maintained by the World Health Organization, SNOMED CT is a general collection of medical terms, LOINC contains codes for medical laboratory observations and RxNorm is used to encode information about drugs.

The same concept may be represented in more than a single ontology. For this reason, it is advisable to exploit a meta-ontology, an ontology that links together entities from different code systems.

UMLS is the most comprehensive medical meta-ontology: it contains 2 million concepts from about 200 ontologies [3].

Even if the 70 % of the descriptions in UMLS are in English, the goal of the project is to create a multilingual knowledge source. If a concept is translated in only a certain ontology, it is possible to discover the relative code in another ontology exploiting the links present in UMLS.

Mapping Strategies and Disambiguation. The problem of discovering entities in a text is known in computational linguistic as Named Entity Recognition (NER). In general, it is possible to exploit rule-based techniques, statistical approaches and dictionary-based lookup to perform this task. In the CDA context, the usage of an ontology is compulsory in order to discover the code related to the discovered entity; rules are only useful to identify dates or physical quantities.

Discovering concepts present in a section and mapping them to the most appropriate UMLS entity is not an easy problem because of the inherent ambiguity of the natural language, the presence of many abbreviations for commonly used medical terms and the fragmented syntax used in medical documents [13].

The most effective approach is to look in the text for sequences of words that match the description of an entity, completely or at least partially. For each section, only the most appropriate categories of entities should be used. For example, if a section of the C-CDA standard contains a list of prescriptions, only the entities that are a drug will be considered.

Disambiguate a concept means selecting the most relevant mapping amount a set of possible candidates. A distance among the different pairs of strings needs to be computed and then the most similar string is selected. If no candidate is clearly winning, it is advisable to avoid the encoding of such concept.

Coreference Resolution. It is not enough identifying key entities. Depending of the type of section other information may be required by the C-CDA standard. As usual it possible that such requirements are not completely fulfilled by the analyzed document.

The problem of grouping together semantically related concepts is called coreference resolution. In a clinical document the entities that are linked to a concept can be identified using a rule-based approach. Such rules need to be carefully designed and are dependent on the section.

For example, after having recognized a drug, it is necessary to look for a dosage and a time period; after a laboratory exam, a result with a physical unit. Given the richness of the C-CDA model, it is necessary to start analyzing the documents that need to be converted in order to look for common patterns that can be mapped to a specific CDA construct.

4 Conclusion

The conversion of medical documents from an unstructured PDF to a XML file following the C-CDA specifications is a challenging task consisting of many different sub problems.

EHR system vendors should consider increasing the amount of information included in the generated PDF files. At least the structure of the document should be present, also for accessibility reasons.

Because of the flexibility of the PDF format, it is possible to include in the same file the CDA version of the document: this is an interesting solution to create a medical document that is easily processable both by humans and machines.

References

1. Aronson, A.R.: Metamap: mapping text to the UMLS metathesaurus. NLM, NIH, DHHS, Bethesda, pp. 1–26 (2006)
2. Bloomberg, D.S., Chen, F.R.: Document image summarization without OCR. In: International Conference on Image Processing, vol. 2, pp. 229–232 (1996)
3. Bodenreider, O.: The unified medical language system (UMLS): integrating bio-medical terminology. Nucleic Acids Res. **32**(1), D267–D270 (2004)
4. Burton, R., Coleman, E., Lipson, D.J., Agres, T., Schwartz, A., Dentzer, S.: Health policy brief: care transitions. Health Affairs (2012). http://www.healthaffairs.org/healthpolicybriefs/brief.php?brief_id=76
5. Carpenter, B., Baldwin, B.: Text Analysis with LingPipe 4. LingPipe Inc., New York (2011)
6. Cavnar, W.B., Trenkle, J.M., et al.: N-gram based text categorization. Ann Arbor MI **48113**(2), 161–175 (1994)
7. Chao, H., Fan, J.: Layout and content extraction for PDF documents. In: Marinai, S., Dengel, A.R. (eds.) DAS 2004. LNCS, vol. 3163, pp. 213–224. Springer, Heidelberg (2004)
8. Dolin, R.H., Alschuler, L., Boyer, S., Beebe, C., Behlen, F.M., Biron, P.V., Shabo, A.: HL7 clinical document architecture, release 2. J. Am. Med. Inform. Assoc. **13**(1), 30–39 (2006)
9. Holzinger, A., Schantl, J., Schroettner, M., Seifert, C., Verspoor, K.: Biomedical text mining: state-of-the-art, open problems and future challenges. In: Holzinger, A., Jurisica, I. (eds.) Interactive Knowledge Discovery and Data Mining in Biomedical Informatics. LNCS, vol. 8401, pp. 271–300. Springer, Heidelberg (2014)
10. International, H.L.S: HL7 implementation guide for CDA release 2: consolidated CDA templates for clinical notes (2014)
11. Kripalani, S., LeFevre, F., Phillips, C.O., Williams, M.V., Basaviah, P., Baker, D.W.: Deficits in communication and information transfer between hospital-based and primary care physicians: Implications for patient safety and continuity of care. JAMA **297**(8), 831–841 (2007)
12. Mao, S., Rosenfeld, A., Kanungo, T.: Document structure analysis algorithms: a literature survey. In: Electronic Imaging 2003, pp. 197–207. International Society for Optics and Photonics (2003)

13. Meystre, S.M., Savova, G.K., Kipper-Schuler, K.C., Hurdle, J.F.: Extracting information from textual documents in the electronic health record: a review of recent research. IMIA Yearb. Med. Inform. **35**, 128–144 (2008)
14. Paskin, N.: Digital object identifier (DOI) system. Encycl. Libr. Inf. Sci. **3**, 1586–1592 (2008)
15. Pustejovsky, J., Stubbs, A.: Natural Language Annotation for Machine Learning. O'Reilly Media, Sebastopol (2012)
16. Savova, G.K., Masanz, J.J., Ogren, P.V., Zheng, J., Sohn, S., Kipper-Schuler, K.C., Chute, C.G.: Mayo clinical text analysis and knowledge extraction system (cTAKES): architecture, component evaluation and applications. J. Am. Med. Inf. Assoc. **17**(5), 507–513 (2010)
17. Trotter, F., Uhlman, D.: Hacking Healthcare: A Guide to Standards, Workflows and Meaningful Use, pp. 172–182. O'Reilly Media, Sebastopol (2011). Chap. 11
18. Trotter, F., Uhlman, D.: Hacking Healthcare: A Guide to Standards, Workflows and Meaningful Use, pp. 144–159. O'Reilly Media, Sebastopol (2011). Chap. 10
19. Xu, H., Stenner, S.P., Doan, S., Johnson, K.B., Waitman, L.R., Denny, J.C.: MedEx: a medication information extraction system for clinical narratives. J. Am. Med. Inform. Assoc. **17**(1), 19–24 (2010)
20. Yimam, S.M., Biemann, C., Majnaric, L., Šabanović, Š., Holzinger, A.: An adaptive annotation approach for biomedical entity and relation recognition. Brain Inform. **3**, 1–12 (2016)

Importance and Quality of Eating Related Photos in Diabetics

Kyriaki Saiti, Martin Macaš[✉], and Lenka Lhotská

Czech Technical University in Prague, Prague, Czech Republic
mmcas@seznam.cz

Abstract. Data are the crucial component of most computer based clinical decision support systems. This review focuses on data for a system which should improve everyday life of diabetics. The aim is to identify issues arising during the process of acquisition of photos of dishes obtained by diabetic patients. Solutions are proposed that will improve the quality of subsequent processing and final conclusions. This research will lead to a proposal of some guidelines that patients should follow when taking the pictures of dishes. For this purpose, a sample of 906 photos from 6 patients including meals and text records of activities was examined carefully in order to extract useful information about how do diabetics chose to record the details, how much and how long do they follow the suggestions. Based on the analysis, representative examples are presented with corresponding suggestions for each case.

Keywords: Data analysis · Data quality · Food · Diabetes · Photos · Meals

1 Introduction

The International Diabetes Federation (IDF) is an umbrella organization of over 230 national diabetes associations in 170 countries and territories. It represents the interests of the growing number of people with diabetes and those at risk. The Federation has been leading the global diabetes community since 1950. According to IDF, 415 million people have diabetes in the world and more than 59.8 million of them are from the EUR Region; by 2040 this will rise to 71.19 million[1]. There were 799,300 cases of diabetes in Czech Republic in 2015. Based on the research, the prevalence of diabetes in adults (20–79 years) is 9.9 %, while the total number of cases of diabetic adult patients (from the same age) are 799.3 (1000s) (see Fig. 1).

Diabetes mellitus (or simply diabetes) is a chronic disease that occurs when the pancreas is no longer able to make insulin (*type 1 diabetes mellitus*), or when the body cannot make good use of the insulin it produces (*type 2 diabetes mellitus*). Both cases lead to an increased blood glucose levels (called hyperglycemia), which if causes diabetic complications such as heart failure, heart attack, lesions

---

[1] http://www.idf.org/who-we-are.

© Springer International Publishing Switzerland 2016
M.E. Renda et al. (Eds.): ITBAM 2016, LNCS 9832, pp. 173–185, 2016.
DOI: 10.1007/978-3-319-43949-5_13

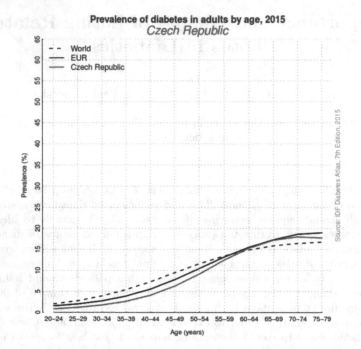

Fig. 1. The figure describes which age groups in the population have the highest proportions of diabetes. The *dotted* line is the distribution of diabetes prevalence by age for the world; the *black* line is the distribution for the region; and the country distribution is plotted in the *red* line. Many middle- and low-income countries have more people under the age of 60 with diabetes compared to the world average. Meanwhile, for high-income countries, a growing population over the age of 60 makes up the largest proportion of diabetes prevalence. (Color figure online) (Source: IDF Diabetes Atlas, 7th Edition, 2015)

of the retina, nerve damage etc. The complexity of diabetes and its effects highly motivates the research on diabetes, where an understanding of the effect of food on glucose levels plays an important role.

Clinical Decision Support Systems (CDSS) can help doctors and patients with the treatment of diabetes. One of CDSS, FEL-Expert system, is a universal diagnostic expert system that is able to provide expert advice, decision or solution recommendation in a particular situation. Figure 2 shows the basic architecture of such an expert system [4]. The first and one of the most crucial steps in the development of such system is data acquisition, information extraction and drawing conclusion. In the light of the acquisition of data by patients, the quality (not only the quantity) of the data can help the experts to increase reliability of the conclusions and simultaneously reduces an unnecessary and bothering effort related to the data acquisition For those reasons, a set of examples was collected, specific quality issues are detected and their solutions are proposed in this paper.

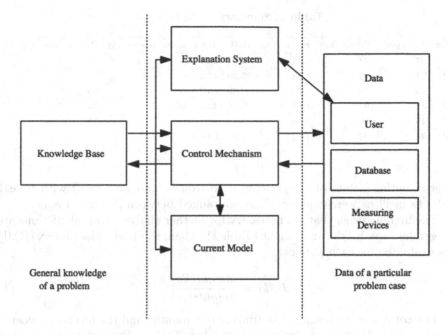

Fig. 2. Block schema of a diagnostic expert system   (Source: Uncertainty processing in FEL-Expert, Lecture notes, [4])

2   Diabetes Mellitus, Carbs and Activity

The monitoring of diabetes includes monitoring of weight loss, exercise and healthy eating (meals that optimally combine all three food categories ratio). The food ingredients are proteins (e.g. meat, fish, cheese, egg, milk), carbohydrates and fat [3]. It is a myth that diabetics should not eat carbohydrates. It used to be advise by physicians before insulin became available, because it was known that carbohydrates increase the blood glucose levels. Today, however, diabetics can enjoy a healthy and balanced diet, which should include carbohydrates [2]. After many years of several investigations and efforts, it becomes clearer than ever how important is to incorporate proper nutrition and exercise in everyday life.

3   Pilot Study

In order to be able to create a knowledge base for a CDSS, we asked patients with *type 1 diabetes mellitus* to record daily consumed meals and snacks, write down the time and take pictures of meals. An information from pictures can provide the calculation of the size of the portions. Further, insulin levels and the values of blood glucose are recorded by sensors within one month. Moreover, it was suggested to patients to record the time that they use bolus insulin. Up

Table 1. Summary of the patients

Sex	Age (yrs)	Weight (kg)	Height (m)	BMI	Total insulin dose (avg)	HBA1c (mmol/mol)
F	27	74	1.55	30.8	41.1	80
M	54	88	1.84	25.99	47.2	39
F	45	70	1.73	23.39	43.2	96
M	47	92	1.89	25.76	43.6	64
M	45	86	1.85	25	40	59
M	35	102	1.92	27.6	53	59

to now, eating habits of six patients (two women and four men) with type 1 diabetes mellitus were captures. The age ranged between 27 and 54 years.

The first elements that are necessary for further analysis and calculations are the gender, age, height and weight (Table 1). Then, the Body Mass Index (BMI) was evaluated for each patient:

$$BMI = \frac{weight(kg)}{height^2(m)} \tag{1}$$

The total average insulin dose (during one month) and the levels of glycated hemoglobin (HBA1c) are also included in the Table 1. By measuring glycated haemoglobin (HbA1c), clinicians are able to get an overall picture about average blood sugar levels. This is important as the higher the HbA1c, the greater the risk of developing diabetes-related complications. The target value for HbA1c of diabetics is 48 mmol/mol (6.5 %) [2]. BMI assigns the persons as underweight (15–18.5), normal weight (healthy, 18.5–25), overweight (25–30) or obese (30–40) [5]. In our sample, one patient is characterized as obese (the first patient), four as overweight, and only one patient was keeping a normal weight. Additionally, all patients have elevated HBA1c levels except one. This first picture makes clear that patients should be motivated to control diabetes in a better and more effective way.

As already mentioned, the process of creating a strong and comprehensive knowledge base will be focused in the project. This obviously needs much more data to be acquired. However first, the quality of data acquisition must be improved to make it possible to get a more useful information from the recordings and pictures. Particularly, a metric systems (e.g.calculation of quantities with linear scales) is missing as well as and the lack of records about activities (gymnastics, walking, etc.) It is noteworthy that while same instructions were given to all patients, almost 6 different versions of the data appeared due different personalities of the patients (e.g. how each one wants to improve his daily life). Moreover, the patients probably differ in the free time that each patient devotes to data acquisition. It was observed that women recorded more pictures than males, which can be observed in the Table 2).

In the following, we will present an analytical explanation of the basic problems which arose in the recording of meals and activities, taking photos and using the metric systems. In any case, we will try to suggest alternatives that

Table 2. Total estimation of number of photos per patient

Sex	Number of photos
F	267
M	163
F	270
M	163
M	70
M	117

should make the process easier for patients and more efficient for the experts who will get a most "powerful weapon" and a more solid foundation for creation of a functional expert system.

3.1   Recording Meals - Text Records

Initially, patients were asked to record meals and snacks that they consumed during one month. Table 3 shows how many records have been gained from each patient. Although instructions were the same for all patients, we noticed that they did not succeed to write down daily meals which probably attributed to lack of time or some other difficulties of recording (e.g. eating out in a restaurant). The complete lack of information for the last patient complicates the analysis of data because the recordings provide additional information that is not "visible" in images. In the case that there is no time to record the necessary details of meals, as may occur when the patient goes into a restaurant or eats during the work, a better proposal would be probably to make voice records using cell phone which is faster and more simple both for the patients and the experts.

Table 3. Number of recordings/total days of records from each patient

Patient	Photos/total days of records
1	30/33
2	30/36
3	32/43
4	19/19
5	30/33
6	0/13

Furthermore, we will try to represent the quality of records content and propose solutions for improvement. We will start with presentation the wrong examples to be avoided. The first information that should be provided is the type of these food or drinks (e.g. type of bread). Most of the times, this information is not evident even in photos but is very important for the analysis of data because there are a few different kinds of beer or bread with different carbohydrate content.

Table 4. Different types of food and carbs

Type of bread	Total carbs (g)
white bread (2 slices serving)	25.30
Rye bread (2 slices serving)	24.51
Mixed grain bread (2 slices serving)	24.13
Dinner roll (1 roll serving)	14.11
Type of rice	Total carbs (g)
White (1 cup of cooked)	44.08
Brown (1 cup of cooked)	44.42
Beer (1 can or bottle)	Total carbs (g)
Regular	12.64
Brown ale	18.00
Light	5.81

Table 4 shows some examples in the content of carbohydrates and demonstrates how important is to know the particular type of the food[2]. For example, if we only see the picture of a glass of beer we can not suppose or understand the type of it. As a result, we should recommend patients to write down the type or take a picture of the bottle of the beer and the ingredients. Another useful information is to record the weight of the food when it is necessary and we can make estimations from the pictures. There are two solutions in order to give the portion size. First, the patient should take a picture of the whole package(in order the grams to be obvious) and refer how much he/she consumed. The other way is to write down the weight (or its estimate). For example, when we see bowl of yogurt it is really difficult to estimate its weight so the patient should add this information in the text records. In Fig. 3 the patient took a picture of the package of the yogurt so we can easily estimate the food portion and the only additional information we need to know is how much of it he/she consumed. In Fig. 4 patient took a picture of a bowl of yogurt and he recorded the whole meal (not the portion size or grams of yogurt in the bowl).

In Fig. 5 the patient recorded details in a better way but not as satisfactory as we need. She recorded the exact grams of meat, we can make an estimation of the amount of potatoes, but it would be useful to know also the type and the amount of vegetables. In summary, we could suggest some simple solutions

---

[2] https://www.fatsecret.com/calories-nutrition.

Fig. 3. Example of yoghurt photo

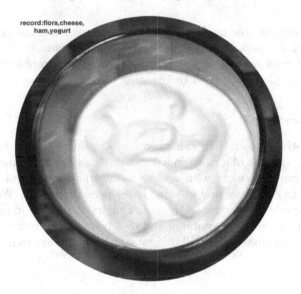

Fig. 4. Another example of yoghurt photo

that will make a difference in the quality of data and make them more reliable. Patients must pay attention to the differences between the different types of the same food (as shown in Table 4), record the quantity (grams, portion size etc.) especially when it is not obvious from the photos (e.g. mashed potatoes; the patient should give details about the recipe and how many potatoes he/she used). In case of lack of time or ability to write down details, voice record should be used. Another important factor is the way of cooking as it has been shown

Fig. 5. Photos from patients(c)

to play an important role on to the calories you get from the consumption of food [1]. Therefore, we advise patients to record the quantity, the type of food and the way of cooking so that together with the photographic material it will make a more complete picture of their eating habits.

3.2 Metric Systems

Obviously, another factor than can play a vital role to the best estimation of portion sizes is addition of a ruler. What can patients use as a ruler and in which way? Apart from the rulers with scales, an alternative suggested to patients are spoons, forks or knives and in some cases pens or pencils. We will initially examine the sample of patients on the use of rulers of all kind in order to discuss the results and notice the advantages and disadvantages of using them. What is more, after the examination we will present some better ways to improve pictures by using rulers.

Table 5. Use of rulers/total days of records from each patient

Patient	Rulers/total days of records
1	59/267
2	14/163
3	129/270
4	13/70
5	60/163
6	16/117

Fig. 6. Photos from patients(d)

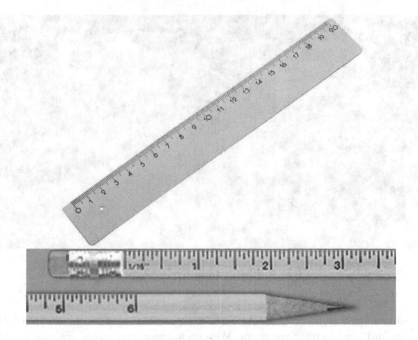

Fig. 7. Suggestions for better estimations

Table 5 shows that the use of rulers or their alternatives (spoon, knife, fork) according to the total registered meals is quite small. We will examine some extended photos and we will introduce some better ways to use them. In most cases, the issue is not just the use of a ruler but also to find the right and

Fig. 8. Photos from patients(e)

Fig. 9. Photos from patients(f)

most convenient way to do that. As a result, we can present some disappointing examples and how to improve them. We can suggest the use of scale as another alternative way of reference ruler. In Fig. 6 the patient decided to use a knife as a ruler, but first of all it is not obvious what the patient wanted to measure. The first suggestion can be to measure separately the ingredients and in different photos to help scientists to make conclusions. Obviously, it is not convenient to use knives (or spoons and forks), because there are many kinds and different scales. It is better prefer a common ruler. However, it is remarkable that no

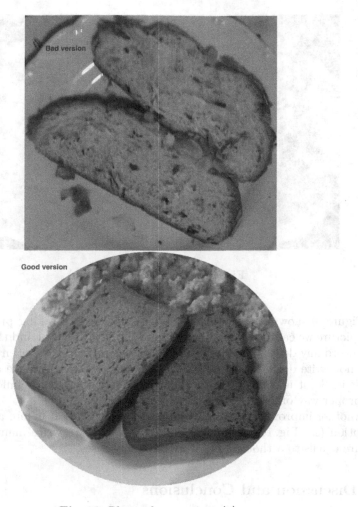

Fig. 10. Photos from patients(g)

patient used this ruler. Another alternative method would be the use of common pencils with scales (Figure 7).

3.3   Quality of Photos

Another important factor is the angle of making photos. It is very useful if we can estimate size of the food (e.g. width, height and thickness of a piece of bread). It is very useful to shoot also the packages of products at a distance from which one can read the ingredients or find the details about the product. A good example of a difference can be found on Figs. 8 and 9. Instead of taking pictures of just the slices of salami, it is better to photograph the packaging of salami and report how many pieces are consumed.

Fig. 11. Photos from patients(h)

Figure 10 shows the difference between the ways of taking pictures. In the first picture we cannot estimate the size of the slice of bread and the patient did not record any details about that, but in the second picture, even if the patient does not write down details about the portion size we can make an estimation about it. So, it is clear that it is important for patients and clinicians to know the proper way of taking pictures to avoid a wasted effort.

Another improvement would be the recording of food before and after consumption (see Fig. 11). Thus, patients do not have to write many details as a picture equals to a thousand of words in many cases.

4   Discussion and Conclusions

The first estimates after the connection of the meal with the glucose levels shown the great need to consult patients about the correct way to combine food and to integrate gymnastics and activities in their everyday lives. These instructions would help patients to manage the disease and will inspire the prevention of it against the sedentary lifestyle that promote from the western society. This project is at the beginning stage but the question of quality, reliability and usefulness of the data is crucial for its future success. This question is also solved by many research groups in recent years. In this review, we suggested solutions useful for the patients but at the same time support experts to reach better conclusions and create a really useful knowledge base for future applications. It is very important for patients to find and realize the real importance of recording the data properly in order to help themselves but also to other patients all over the world.

Acknowledgement. This research is supported by the AZV MZ CR project No. 15-25710A "Individual dynamics of glycaemia excursions identification in diabetic patients to improve self managing procedures influencing insulin dosage" and CVUT grant No. SGS16/231/OHK3/3T/13 "Support of interactive approaches to biomedical data acquisition and processing".

References

1. Birlouez-Aragon, I., Saavedra, G., Tessier, F.J., Galinier, A., Ait-Ameur, L., Lacoste, F., Niamba, C.N., Alt, N., Somoza, V., Lecerf, J.M.: A diet based on high-heat-treated foods promotes risk factors for diabetes mellitus and cardiovascular diseases. Am. J. Clin. Nutr. **91**(5), 1220–1226 (2010)
2. Holt, R.I., Cockram, C., Flyvbjerg, A., Goldstein, B.J.: Textbook of Diabetes. Wiley, Oxford (2011)
3. Kawamura, T.: The importance of carbohydrate counting in the treatment of children with diabetes. Pediatr. Diabetes **8**(s6), 57–62 (2007)
4. Lhotska, L., Marik, V., Vlcek, T.: Medical applications of enhanced rule-based expert systems. Int. J. Med. Inform. **63**(1), 61–75 (2001)
5. Schwab, K.O., Doerfer, J., Naeke, A., Rohrer, T., Wiemann, D., Marg, W., Hofer, S.E., Holl, R.W.: Influence of food intake, age, gender, HbA1c, and BMI levels on plasma cholesterol in 29 979 children and adolescents with type 1 diabetes-reference data from the german diabetes documentation and quality management system (dpv). Pediatr. Diabetes **10**(3), 184–192 (2009)

Univariate Analysis of Prenatal Risk Factors for Low Umbilical Cord Artery pH at Birth

Ibrahim Abou Khashabh[1], Václav Chudáček[2(✉)], and Michal Huptych[2]

[1] Department of Cybernetics, Faculty of Electrical Engineering,
Czech Technical University in Prague, Prague, Czech Republic
[2] Czech Institute of Informatics, Robotics and Cybernetics,
Czech Technical University in Prague, Prague, Czech Republic
vaclav.chudacek@cvut.cz

Abstract. Objective: To identify potential risk factors for low umbilical cord artery pH in term, singleton pregnancies. Methods: Retrospective case-control study. Cases were deliveries characterized by umbilical cord artery $pH \leq 7.05$. Controls were with no sign of hypoxia. Results: In the database of 10637 deliveries, collected between 2014 and 2015 at the University Hospital in Brno delivery ward, we identified 99 cases. Univariate analysis of clinical features was performed. The following risk-factors were associated with low pH: the length of the first stage (odds ratio (OR) 1.40 (95 % CI 1.04–1.89)) and the length of the second stage of labor (OR 2.86 (95 % CI 1.70–4.81)), primipara (OR 2.99 (1.90–4.71)) and meconium stained fluid (OR 1.60 (1.07–2.38)). Conclusion: Among the risk factors that increase the chance of low umbilical cord artery pH at term, we identified: excessive length of the first and second stage of labour, parity, and meconium stained fluid.

1 Introduction

The main focus of intrapartum cardiotocography (CTG) evaluation is on revealing possible occurrences of fetal hypoxia during delivery. Since CTG traces are not evaluated in a vacuum, it is of great importance to be aware of clinical features related to low pH outcomes in the newborn cord artery blood. Although many clinical factors have been documented in the literature as increasing the risk for mother and child during delivery, no study has attempted to connect clinical features with the objective outcome in form of umbilical cord artery pH.

In previous works, the main interest focused on risk factors related to Caesarean section. Gareen et al. [7] investigated the relation between maternal age and rate of Caesarean section. Patel et al. [13] published a study on a population of over 12 000 deliveries describing in great detail the relation between various risk factors and Caesarean section. Verhoeven et al. [18] investigated factors leading to Caesarean section in multiparous women.

Other studies have also investigated risk factors influencing birth outcomes, e.g. the study by Martius et al. [12] focusing on pre-term delivery risk factors

© Springer International Publishing Switzerland 2016
M.E. Renda et al. (Eds.): ITBAM 2016, LNCS 9832, pp. 186–191, 2016.
DOI: 10.1007/978-3-319-43949-5_14

for various degrees of prematurity, and on an assessment of factors contributing to a prolonged labor, e.g. Szal [16].

When searching for relevant research papers, we used the phrase "risk factors for birth asphyxia". This yielded 20 results, out of which only four seemed relevant to the study at hand [1,3,8,9]. We should note here that in both cases the asphyxia was related to the Apgar score – which is a subjective measure and is considered to be only a crude descriptor of asphyxia [10]. A retrospective study by Aslam et al. [1] featured a database with an equal distribution of cases and controls (ration 123 : 117), and found that primiparous women and those with pre-eclampsia had a significantly (p < 0.01) greater risk of a low Apgar score. A study by Chiabi et al. [3] found the following antepartum risk factors: place of antenatal visit, malaria during pregnancy, and preeclampsia. Intrapartum risks included: prolonged labor, stationary labor, and term prolonged rupture of membranes. A study from Uganda on antenatal and intrapartum risks by Kaye [8] reported anemia, pre-eclampsia, meconium stained fluid and low birth weight among the most significant risks with respect to outcome assessed by the Apgar score. In an interesting work of Landfors [9] meconium was highlighted together with several CTG-related parameters as the factors increasing the risk of asphyxia (again assessed based on Apgar score) the most.

In our previous work we have focused on evaluations of CTG recordings and their relation to pH c.f. [4,15], and on evaluation of decision-making processes of expert obstetricians [21] using the CTG recording database available at the University Hospital in Brno [5]. Since year 2014 we have also been collecting structured clinical data on all deliveries. Preliminary results of our analysis are presented bellow.

2   Methods

Our study is a data-driven retrospective exploratory analysis utilizing the University Hospital in Brno database consisting of clinical data as well as the CTG recordings, which are in part published as an open-access database [5]. After a previous attempt to understand in depth the individual relations in the data [11] we focus here on a detailed univariate analysis of the collected features.

2.1   Database

The data were collected using our own data collection system, DeliveryBook, at the delivery ward of the University Hospital in Brno from January 2014 to December 2015. More than 100 clinically relevant parameters about mother, delivery and newborn are stored in a structured way. The collection and use of the data was approved by the ethics committee at the University Hospital in Brno. All data are fully anonymized. The database consists of 12274 recordings, but in order to allow straightforward interpretation we have used the following conditions in order to homogenize the final set: arterial pH is available; gestational age ≤37 weeks; singleton pregnancy; no known congenital diseases. This yielded in total 10637 deliveries for analysis.

Table 1. A collection of the most interesting results from a univariate analysis of the features. pH, Apgar and SC are outcome measures, while all the other features represent knowledge or an action known prior to delivery. Entonox and epidural analgesia are medications given during the labor.

	Cases – $pH \leq 7.05$			Controls – $pH > 7.05$				
	#	prct (%)	mean (std)	#	prct (%)	mean (std)	OR (95 % CI)	p-value
pH ≤ 7.05	99	0.93	6.99 (0.09)	10523	99.07	7.29 (0.08)	-	-
Apgar score 5 min	203	1.91	6.50 (0.88)	10400	98.09	9.65 (0.59)	24.68 (15.49 - 39.31)	<0.001
Sectio Caesarea	2110	20.30	1.00 (-)	8285	79.70	0.00 (-)	1.58 (1.02 - 2.45)	<0.001
Induced delivery	2198	20.69	1.00 (-)	8424	79.30	0.00 (-)	1.44 (0.92 - 2.25)	<0.001
Entonox	144	1.36	1.00 (-)	10478	98.64	0.00 (-)	2.30 (0.72 - 7.35)	<0.001
Epidural analgesia	1915	18.03	1.00 (-)	8707	81.97	0.00 (-)	1.01 (0.60 - 1.69)	1.000
Ist stage (> 360 min)	1807	18.40	440.15 (64.77)	8015	81.60	226.22 (73.79)	1.36 (0.84 - 2.20)	<0.001
IInd stage (> 30 min)	884	10.11	54.12 (20.60)	7862	89.89	11.04 (7.18)	2.86 (1.70 - 4.81)	<0.001
Parity (< 2)	5299	49.99	1.00 (0.00)	5302	50.01	2.34 (0.78)	2.99 (1.90 - 4.71)	<0.001
Sex (Male)	5468	51.53	1.00 (-)	5143	48.47	2.00 (-)	0.69 (0.46 - 1.03)	0.001
O100 – hypertension	533	5.02	-	10089	94.98	-	1.45 (0.67 - 3.13)	<0.001
O140 – preeclampsia	114	1.07	-	10508	98.93	-	1.92 (0.47 - 7.87)	0.008
O365 – IUGR	375	3.53	-	10247	96.47	-	1.78 (0.77 - 4.08)	<0.001
O681 – meconium	784	7.38	-	9838	92.62	-	2.09 (1.18 - 3.69)	<0.001
D650 – defibrination syndrome	258	2.43	-	10364	97.57	-	2.63 (1.14 - 6.06)	<0.001
D695 – secondary thrombocy-topenia	36	0.34	-	10586	99.66	-	6.36 (1.51 - 26.85)	<0.001

2.2   Features

A total of 107 features were analysed all of them falling into one of the following categories: (a) detected prior to going into labour (e.g. parity, sex of the fetus, induction of labour, some diagnosis related to mother and pregnancy) (b) occurring within the delivery period (e.g. interventions, length of delivery stages, medications, diagnosis related to delivery) (c) features known after delivery – outcome measures (e.g. Apgar score, admittance to the NICU, seizures or intubation).

2.3 Statistical Evaluation

The relation between each feature and the pH outcome was evaluated using univariate logistic regression to obtain odds ratios, their 95 % confidence intervals (CI), and two-sided p-values, in addition to basic descriptive statistics.

An odds ratio (OR) is a measure of the association between an exposure and an outcome. It answers the question - how much of a difference does the presence of the risk factor have on your chances of getting the disease [14]? Odds ratios are most commonly used in case-control studies. However they can also be used in cross-sectional and cohort study designs [17]. Additional information on the OR can be found in the statistical notes by Bland [2], with arguments supporting the OR over relative risk (RR) estimation. In our case, the difference between OR and RR is anyway negligible, since the event rate of pathological cases is very low, so $OR \simeq RR$ [14]. All computations were done in Matlab 2015a.

3 Results

All results are shown in Table 1. The main parameter defining the two groups (cases and controls) of newborns was the umbilical artery pH (set to 7.05) – a common set-up value in the studies using pH c.f. e.g. [20,22].

4 Discussion

Several features related to low pH outcomes were identified. However some uncertainties on the side of the database and the interpretation of the results will have to be clarified before we move on to the multivariate analysis.

First, although the size of the database is promising, the main concern is the size of the subsets representing individual features. Especially for diagnostic features where ICD-10 codes are used, many subsets contain only tens of cases, which is insufficient to allow conclusions to be drawn with high confidence.

Second, although all features, with the exception of epidural analgesia, are deemed significant with $p < 0.01$, it is necessary to interpret carefully the clinical relevance of all OR that span over $OR = 1$.

According to our review of the available literature, no studies have related the clinical features known before and during delivery to the outcome in the form of the pH value. Although some studies e.g. by Dani [6] suggest the link between low pH and an adverse outcome is only weak, there are other studies, e.g. by Yeh [19] showing a significant increase in the risk of neurological impairment with $pH \leq 7.1$. Thus, irrespective of the controversy surrounding pH as an outcome measure it should be an imperative to avoid a low pH value.

Our study highlights features related to low pH. Although many of the parameters are known to increase the risk for the baby our data driven analysis gives additional evidence with respect to low pH.

The ultimate goal in utilizing data available from the database is to marry the clinical data analysis with CTG/FHR analysis in order to provide comprehensive decision support during delivery. The study presented here is just an initial investigation into the database.

5   Conclusion

An analysis of risk factors for low umbilical cord artery pH was undertaken and identified several interesting features related to parity, sex of the fetus, induced labour, and also those related to specific diagnosis codes related with the delivery – e.g. meconium staining (O681) and defibrination syndrome (D650).

Acknowledgment. This work was supported by SGS grant SGS16/231/ OHK3/3T/13 of the CTU in Prague and by Czech Science Foundation Agency project 14-28462P Statistical methods of intrapartum CTG signal processing in the context of clinical information. In addition we would like to express our thanks to the team at the University Hospital in Brno, in particular to Dr. Janků and Dr. Hruban.

References

1. Aslam, H.M., Saleem, S., Afzal, R., Iqbal, U., Saleem, S.M., Shaikh, M.W.A., Shahid, N.: Risk factors of birth asphyxia. Ital. J. Pediatr. **40**(1), 1–9 (2014)
2. Bland, J.M., Altman, D.G.: Statistics notes. The odds ratio. BMJ **320**(7247), 1468 (2000)
3. Chiabi, A., Nguefack, S., Evelyne, M., Nodem, S., Mbuagbaw, L., Mbonda, E., Tchokoteu, P.-F., Anderson, D.: Risk factors for birth asphyxia in an urban health facility in cameroon. Iran. J. Child Neurol. **7**(3), 46–54 (2013)
4. Chudáček, V., Spilka, J., Huptych, M., Georgoulas, G., Lhotská, L., Stylios, M., Koucký, C., Janků, P.: Linear and non-linear features for intrapartum cardiotocography evaluation. In: Computers in Cardiology, vol. 35 (2010)
5. Chudáček, V., Spilka, J., Burša, M., Janků, P., Hruban, L., Huptych, M., Lhotská, L.: Open access intrapartum CTG database. BMC Pregnancy Childbirth **14**(1), 16 (2014)
6. Dani, C., Bresci, C., Berti, E., Lori, S., Di Tommaso, M., Pratesi, S.: Short term outcome of term newborns with unexpected umbilical cord arterial pH between 7.000 and 7.100. Early Hum. Dev. **89**(12), 1037–1040 (2013)
7. Gareen, I.F., Morgenstern, H., Greenland, S., Gifford, D.S.: Explaining the association of maternal age with cesarean delivery for nulliparous and parous women. J. Clin. Epidemiol. **56**(11), 1100–1110 (2003)
8. Kaye, D.: Antenatal and intrapartum risk factors for birth asphyxia among emergency obstetric referrals in mulago hospital, kampala, uganda. East Afr. Med. J. **80**(3), 140–143 (2004)
9. Ladfors, L., Thiringer, K., Niklasson, A., Odeback, A., Thornberg, E.: Influence of maternal, obstetric and fetal risk factors on the prevalence of birth asphyxia at term in a swedish urban population. Acta Obstet. Gynecol. Scand. **81**(10), 909–917 (2002)
10. Leuthner, S.R., et al.: Low apgar scores and the definition of birth asphyxia. Pediatr. Clin. North Am. **51**(3), 737–745 (2004)
11. Burša, M., Lhotská, L., Chudáček, V., Spilka, J., Janků, P., Hruban, L.: Information retrieval from hospital information system: increasing effectivity using swarm intelligence. J. Appl. Logic **13**(2), 126–137 (2015)

12. Martius, J.A., Steck, T., Oehler, M.K., Wulf, K.H.: Risk factors associated with preterm (< 37+ 0 weeks) and earlypreterm birth (< 32 + 0 weeks): univariate and multivariate analysis of 106345 singleton births from the 1994 statewide perinatal survey of bavaria. Eur. J. Obstet. Gynecol. Reprod. Biol. **80**(2), 183–189 (1998)

13. Patel, R.R., Peters, T.J., Murphy, D.J., ALSPAC Study Team, et al.: Prenatal risk factors for caesarean section. Analyses of the ALSPAC cohort of 12 944 women in england. Int. J. Epidemiol. **34**(2), 353–367 (2005)

14. Schmidt, C.O., Kohlmann, T.: When to use the odds ratio or the relative risk? Int. J. Publ. Health **53**(3), 165–167 (2008)

15. Spilka, J., Chudáček, V., Janků, P., Hruban, L., Burša, M., Huptych, M., Zach, L., Lhotská, L.: Analysis of obstetricians' decision making on CTG recordings. J. Biomed. Inform. **51**, 72–79 (2014)

16. Szal, S.E., Croughan-Minihane, M.S., Kilpatrick, S.J.: Effect of magnesium prophylaxis and preeclampsia on the duration of labor. Am. J. Obstet. Gynecol. **180**(6), 1475–1479 (1999)

17. Szumilas, M.: Explaining odds ratios. J. Can. Acad. Child Adolesc. Psychiatry **19**, 227–229 (2010)

18. Verhoeven, C.J., van Uytrecht, C.T., Porath, M.M., Mol, B.W.J.: Risk factors for cesarean delivery following labor induction in multiparous women. J. Pregnancy **2013** (2013). http://dx.doi.org/10.1155/2013/820892

19. Yeh, P., Emary, K., Impey, L.: The relationship between umbilical cord arterial pH and serious adverse neonatal outcome: analysis of 51,519 consecutive validated samples. BJOG **119**(7), 824–831 (2012)

20. Doret, M., Spilka, J., Chudáček, V., Gonçalves, P., Abry, P.: Fractal analysis and hurst parameter for intrapartum fetal heart rate variability analysis: a versatile alternative to frequency bands and LF/HF ratio. PLoS ONE **10**(8), e0136661 (2015)

21. Hruban, L., Spilka, J., Chudáček, V., Janků, P., Huptych, M., et al.: Agreement on CTG intrapartum recordings between expert obstetricians. J. Eval. Clin. Pract. **21**(4), 694–702 (2015)

22. Karvelis, P., Spilka, J., Georgoulas, G., Chudáček, V., Stylios, C.D., Lhotská, L.: Combining latent class analysis labeling with multiclass approach for fetal heart rate categorization. Physiol. Meas. **36**(5), 1001–1024 (2015)

Applying Ant-Inspired Methods in Childbirth Asphyxia Prediction

Miroslav Bursa$^{(\boxtimes)}$ and Lenka Lhotska

Czech Institute of Informatics, Robotics and Cybernetics,
Czech Technical University in Prague, EU, Prague, Czech Republic
{miroslav.bursa,lenka.lhotska}@cvut.cz

Keywords: Data mining · Textual data processing · Data visualization · Ant colony optimization · Hospital information system · Decision tree induction

1 Introduction

In the today's world we witness an impact of the 'Big data' phenomenon. Although there are many definitions and different scientists view the problem from their perspectives (web, IoT, smartphones, security, GIS, HIS, cloud systems, networks, ...), there is still need for efficient, robust and scalable algorithms that ease processing of such data. This paper deals with data that might not fit into the *Big data* definitions, however the amount of the work needed even for such smaller data-mining problem, is enormous. The main reason is the heterogeneity of the data, multiple variants and synonyms in the terminology used, natural language processing caveats and many errors, omissions and typos (spelling errors).

In this paper we target the process of creating a decision support system induced from data, signals and information contained within a hospital facility (facilities). We do not describe the long process of integration of different sources and the long way needed to gain access to various systems on multiple levels. The aim of the resulting decision support system is quite straightforward: to help the medical expert in decision on the severeness of child asphyxia during a childbirth. Based on the evaluation, the expert usually decides whether the delivery should continue in a natural way, or a Caesarian section should be indicated.

In the last two decades, many advances in the computer sciences have been based on the observation and emulation of processes of the natural world. The origins of *bioinspired informatics* can be traced to the development of perceptrons and artificial life, which tried to reproduce the mental processes of the brain and biogenesis respectively, in a computer environment [1]. Bioinspired informatics also focuses on observing how the nature solves situations that are similar to engineering problems we face.

Nature inspired metaheuristics offer fast and robust solutions in many fields (graph algorithms, feature selection, optimization, clustering, various NP and

© Springer International Publishing Switzerland 2016
M.E. Renda et al. (Eds.): ITBAM 2016, LNCS 9832, pp. 192–207, 2016.
DOI: 10.1007/978-3-319-43949-5_15

NP-Hard problems, etc.). Stochastic nature inspired metaheuristics have interesting properties that make them suitable to be used in data mining, data clustering and other application areas.

By studying the behavior in nature, namely ant colonies, we can easily find many inspiring concepts for various metaheuristics. The high count of simple agents and the decentralized approach to task coordination in the species studied indicates that ant colonies show a high degree of parallelism, self-organization and robustness.

We hereby concentrated on the methods that are inspired by the processes that are viable in the nature. These are widely known and utilized (such as artificial neural networks, genetic algorithms and many other). The advantage of such methods lies in robustness, decentralization, fault recovery and tolerance and their use is justified by the long and uncompromising process of evolution. Moreover, in hybridization with classical methods, the nature-inspired methods offer various techniques, that can favourably increase speed and robustness of the classical methods.

1.1 Motivation and Clinical View

Severe asphyxia during the process of childbirth can lead to various complications, resulting into serious brain damage of the neonate as the brain is the most sensitive to oxygen deprivation. If the doctors were able to detect early stages of asphyxia, Caesarian section might be indicated in time in order to speed up the whole process and reduce the time the newborn spends in oxygen insufficiency. However, during the delivery, certain (but not severe) asphyxias are perfectly normal and natural part of the childbirth process.

On the other hand, the doctors aim to avoid unnecessary Caesarian sections as much as possible. In fact, they perform multiobjective optimization, while minimizing the overall risk. This is, however, hard to evaluate quantitatively and the doctors decide based on their experience. The detection and decision process is of course much more complicated and the prediction requires long-term praxis of the expert and is usually based on the cardiotocography signal (CTG) that is being monitored during the delivery. Many studies showed that CTG signal itself is not sufficient and supplemental clinical information is needed for making the correct decision – such as APGAR score, biochemical markers (pH value, base excess, base deficit), family and mother anamneses, risk factors and others. These information are often available in the hospital information system. Sadly, almost never in a format that allows direct or easy processing

The overall aim of this work is to ease the work of obstetricians and to help them improve their decision to better preserve the health of the mother and the neonate. In the paper we describe the technical aspects of the steps needed to create such decision support system.

2   Methodology

2.1   Approaches Used

This work presents the use of paradigms, inspired by the behavior of ant colonies, for various tasks in the whole process of medical information mining from a hospital information system. We have used the aforementioned paradigms for clustering similar text entities together to ease the human task of visual structural discovery in a free text. Further, in the phase of cardiotocography data classification, we have utilized an ant-colony inspired approach for decision tree induction.

A vast number of nature inspired methods are studied and developed in present. One category is represented by methods, that are inspired by the behavior of ant colonies. These methods have been applied to many problems (often NP-hard). Review can be seen in [2,6]. We concentrate on the state-of-the-art nature methods inspired by the social behavior of insect communities, by the swarm intelligence, brain processes and other real nature processes.

Also, ontologies have become an important means for structuring knowledge and building knowledge-intensive systems. For this purpose, efforts have been made to facilitate the ontology engineering process, in particular the acquisition of ontologies from texts.

The accuracy for relation extraction in journal text is typically about 60 % [7]. To reach a perfect accuracy in text mining is nearly impossible due to errors and duplications in the source text. Even when linguists were hired to label text for an automated extractor, the inter-linguist disparity is about 30 %. The best results are obtained via an automated processing supervised by a human [10]. We have to admit, that although our objective to make the process fully automated, the expert intervention and analysis is still inevitable and is not negligible.

In the process of modeling the data, the aim is to obtain a model (classifier, decision tree) that represents the information within the data and to serve as a part of the knowledge base. In this process we have (among others) utilized an ant-colony inspired approach. This approach was compared with other approaches (such as artificial neural networks, Bayes classifier, Kernel methods and other methods for decision tree induction [3].

2.2   Problem Outline

This and related works consisted of the following main phases: (1) Integration of various heterogeneous sources, (2) mining information from free text records of the information system, (3) processing of the CTG data, (4) complementing the CTG data with information from the information system, (5) creating a model of the data (designing classifier or decision tree) and (6) decision support system for asphyxia prediction. This paper describes mainly the phases (2) and (5).

In the text-mining part of this work the main task was to get a simplification of the textual data. It was important to use computer aid, as it was too exhausting for human operators (experts). First, basic statistical evaluation and preprocessing has been made in order to get a higher-level overview of the data stored within the hospital information system. Later, we have utilized a visualization approach that significantly eased rule mining and automated information retrieval. The rules form an important part of the decision support system being designed. The proposed approach brings an important benefit in reduction of time needed for processing free text that was loosely structured (formatting is performed by the newline characters mainly).

Several techniques to extract knowledge from raw text data have been already developed. These techniques have various and multiple origins: some result from the statistical analysis of the data, regression analysis, decision trees, etc.; some originate from the branch of artificial intelligence such as the expert systems, intelligent agent systems, fuzzy logic, etc.

In data mining process and knowledge discovery, exact or heuristic methods can be used. Exact methods provide better results, but require a lot of computational power. But sometimes more robust results may be needed – the exact methods can get over-trained – the method provides very good results on the training data set, but performs poorly on the testing data set (unknown data). In such case the approximation techniques can be used. They yield only approximate (suboptimal) solution (with acceptable error level) which may provide rather high generalization ability. The appropriateness of the use of approximate methods must be carefully considered.

Problem Statement. This paper addressed the following (technical) problems. First, data from the database had to be retrieved and categorized. The problem itself was not in the DB access or structure, but rather in the fact that the data are stored in free text with no structure. Or, we might say – loose structure – created by new lines. But this could not be relied upon as the recordings were hugely affected by the personality of the person who entered the text. The first task was to extract, load and transform (with anonymization step) these data in relevant database system and to remove obvious errors. In this task we utilized visualization to ease the effort.

The second problem described in this paper is the following. After we have obtained CTG signal data with relevant information from the information system, we had to correctly classify the patient records and data. In the second part of the paper we addressed this problem by running an experiment using different classifiers. The data preprocessing, namely feature extraction is described in [4].

The methods for CTG signal pre-processing are not part of this paper, as these are available together with the CTG signal database made available for comparative studies[1]. The study on the expert obstetrician's agreement has been also published [5,9,11]. We therefore encourage other researchers to take advantage of this free database available.

---

[1] The open-access database [4] is freely available at the following link: http://www.physionet.org/physiobank/database/ctu-uhb-ctgdb/.

3   Dataset Overview and Basic Preview

The whole textual dataset to be processed consisted of approximately 50 000 to 120 000 records × cca 20 attributes (these were distributed in various RDB tables). It is an export of a period over 12 years of the obstetrician department of a hospital, featuring about 5.000 childbirths a year. The number of records is not constant for each patient and depends on various parameters and history of pregnancy (pregnancies). Each attribute item contained about 800 to 1 500 characters of text (diagnoses, patient state, anamneses, medications, notes, etc.).

These textual corpses reflect the way the doctors used to work with typewriter machines before deployment of the information system. The paper forms were used as a base to create a basic information system (rather an electronically backed-up paper records). The main problem is the non-structured text without any checks, high variability among the operators and terminology, high rate of spelling errors.

The paper documentation is still used nowadays and the paper documents are typed into the system during night shifts (by a nurse usually). This also increases number of errors in the data.

3.1   Initial Analysis

The first information (apart from visual inspection) has been obtained by performing a statistical analysis of n-grams. The most frequent n-grams have been analyzed to find the most important and descriptive indicators. In order to automatically retrieve important features from the text we selected an approach using regular expressions (RE). These are very flexible and can rather efficiently search in the input text. The problem was the design of the expressions and their verification. Apart from the n-gram analysis the need to visualize the data emerged in order to make the RE design more effective. We have created a tool (in Java, using JUNG[2] framework) to visualize the attributes in a way that helps the experts and linguists to obtain an overview of the problem.

3.2   Practical Problems

First, we have noted that the data contain many errors and typos. See Table 1. You can see that the word *Mesocaine* has more than 100 variants (we have omitted some ambiguous variants, variants where second word was merged and variants with low frequency). As a first step to improve the visualization we have colorized similar words with the same color (i.e.: The Mesocaine-related words are orange).

3.3   Graphical Visualization

According to the feedback, this helped the experts with an orientation within the structure of the texts. But still, the visualization was not that representative.

---

[2] http://jung.sourceforge.net/.

Table 1. Outline of the practical problems: variants and misspellings of the the most prominent words. Only the unambiguous variants have been displayed.

Base word	No variants	Variants incl. typos
Mesocain	98	esocain, maesocain, masocain, masocaine, measocain, mecocain, mecocaine, mecocainu, mecosain, meocain, meocaine, meoscain, meoscaine, meoskain, meosocain, mersocain, merzocain, mesacain, mesaocain, mescain, mescoain, mescocain, mesdocain, mesecain, meseocain, mesiakin, mesicain, mesicainu, meskoain, mesoacain, mesoacaine, mesoacainu, mesoacin, mesoacine, mesoacinu, mesoain, mesoakin, mesoc, mesocai, mesocaien, mesocail, mesocaim, mesocain, mesocaine, mesocainl, mesocainn, mesocainu, mesocaion, mesocaionn, mesocalin, mesocan, mesocani, mesocaoin, mesocaon, mesocasin, mesocaun, mesoccain, mesocein, mesocian, mesociane, mesocianu, mesocien, mesocin, mesocinu, mesocsin, mesoicain, mesokail, mesokaim, mesokain, mesokaine, mesokainu, mesokeain, mesoscain, mesosocain, mesovain, mesovcain, mespcain, messocain, mezcain, mezoacin, mezoakin, mezocail, mezocain, mezocaine, mezocan, mezocian, mezocin, mezokain, mezokainu, mezokian, mezolain, mezovain, mosocain, mrsocain, msocain, msokain, mwesocain, mwsocain
Epiduralis	66	apiduralni, edpiduralni, eiduralis, eiduralni, eipduralni, eoiduralni, epdiduralni, epdiralni, epdirual, epdirualis, epdirualni, epdiural, epdiuralni, epduralni, epi, epid, epidaralni, epidiral, epidiralni, epidral, epidralni, epidrualni, epidualni, epiduaralis, epiduarlni, epiduiralni, epidur, epidural, epiduralbi, epidurali, epiduralis, epiduralmi, epiduraln, epiduralna, epiduralni, epiduralnian, epiduralnii, epidurani, epiduranian, epidurialni, epidurilni, epidurla, epidurlani, epidurlni, epidutal, epidutalni, epiedural, epiidural, epioduralni, epipdrualni, epirual, epirualni, epirudal, epirudalni, epirural, epiruralni, episuralni, epiuralni, epoduralni, epoiduralni, eppiduralis, epuduralni, espiduralni, eupiduralni, ewpiduralni, piduralni
Analgesia	59	aanigesio, ahalegsio, alalgezie, anagesia, anagesie, anagesio, anagezie, anaglesie, anaglezie, analagesia, analagezie, analegesie, analegezie, analegsio, analesio, analg, analgaesia, analgazie, analgecsie, analgeie, analgeio, analgeise, analgeize, analgerzie, analgesoi, analgooi, analgesia, analgesie, analgesien, analgesio, analgesis, analgesui, analgeuie, analgez, analgeza, analgezeie, analgezi, analgezia, analgezie, analgezii, analgezio, analgezis, analgizie, ananlgezie, anaulgesio, anealgezie, anelgesie, anelgezie, anlagezie, anlgeise, anlgesie, anlgezie, anmalgesie, asnalgesie, eanalgesio, enekgesie, nalgesie, nalgesio, nalgezie
Anesthesia	37	...

To get a notion what we are looking for, we let the users organize the graphs according to their needs and with the aim to replicate their organization in an automated way. We used two different approaches to represent a numeric literal: first, we have bound the literal to the preceding word, second, the literal has been left alone as a node.

Number literal (a wildcard) had the highest potency, as many quantitative measures are contained in the data (medication dose, etc.). Therefore it has been fixed to the following literal, splitting into multiple vertices (i.e. a sequence *mesocain 10 mL* becomes two vertices – *mesocaine_NUMBER_* and *mL*). This allowed to organize the chart visualization in more logical manner, see Figs. 1 and 2. However, the problem was that sometimes the _NUMBER_ literal was bound to inappropriate word (as the doctors can write: *10 mL mesocaine* or *mesocaine 10 mL*. This is not an easy task for the automated analysis to decide. Therefore we have also used the non-prefixed version for visualization, such as in Fig. 4.

A problem is that the unit might be omitted, the number literal might be prefixed or suffixed to the object. Time needed to organize such graph was about 5–15 min. Another problem is that the transition graph contained loops, complicating the manual organization.

We have observed that the vertices in a human-only organization were (usually) organized depending on the position in the text (distance from the sentence start) as they have the highest potency (which is also reflected by the vertex diameter in the visualized graphs).

Visualization Described. In this paper a visualization in the form of *transition graphs* that preserves the structure of the sentence has been used. Moreover we have displayed the prevalence of words and their connectedness within the corpus. These graphs were created for individual attributes. An attribute consisted of multiple records in form of a sentence. By *sentence* we hereby mean a sequence of literals (words), not a sentence in a linguistic form. The records were reduced – unnecessary words (such as verbs *is, are*) are omitted. In this paper, only the attribute describing the anesthesia during deliveries is visualized, as it is the simplest one allows to be easily visualized here (and only about 5 % of the record is displayed, otherwise it renders as a black stain).

Vertices of the transition graph represent the words (separated by blank characters, commas or punctuations) in the records. For each word (single or multiple occurrence) a vertex is created and its potency (number of occurrences is counted). For example, the words *mesocaine, anesthetics, not, mL* form a vertex each. For a number (i.e. sequence of digits) a special literal _NUMBER_ is used.

Edges of the graph are created from single records (from the sentences retrieved). For example the sentence "*mesocaine 10 mL*" would add edges from vertex *mesocaine* to vertex _NUMBER_ and from vertex _NUMBER_ to the vertex *mL* (or the edge count is increased in case it already exists).

For all records, the number (count) of the edges is also useful. It provides an overview on the inherent structure of the data – the most often word connections and sentence flow.

4 Dataset Visualization

We have visualized the data in form of a directed graph where each word (literal) represents a vertice and two successive literals form an edge. This way we are able to visualize the most common words and n-grams in the corpus.

For the visualization we have utilized the following widgets. The vertices (literals) are represented as filled circle, the diameter reflects its potency (frequency). Edges represent transition states between literals (the sequence of 2 subsequent words in a sentence/record); edge thickness reflects the transition rate (probability) of the edge. The same holds for all subsequent figures showing the transition graph, only a different visualization approach has been used. It is clear, that human interpretation and analysis of the textual data is very fatiguing, therefore any computer aid is highly welcome.

4.1 Human Organization

In order to use an automated approach, we had to acquire an idea about what we were searching for. Therefore humans were asked to organize the structure according to expert's needs keeping in mind their future use for designing the regular expressions. We have repeated the process using various attributes. An example of human organization is available in Figs. 4 and 6. Note that one of the experts (Fig. 6) decided to overlap the nodes one above the other. This helped to (visually) merge the edges and better revealed the structure of the record. The second (Fig. 4) carefully positioned the nodes so that the labels remained perfectly visible.

4.2 Automated Organization

The first automated attempt was a simple positioning based on the word distance from the sentence start. Although it looked impressively, the same words were mispositioned in the horizontal axis (i.e. the words related to *mesocain* were placed in multiple columns). Therefore we decided to solve the issue by clustering similar vertices together.

We have tested a simple k-means approach to cluster the words (vertices). The k-means algorithm quite quickly combined similar vertices together, however the main problem was to keep the clusters apart from each other – as the algorithm had tendency to compact the whole graph into a very small object.

Another approach implemented was the self-organizing map (Kohonen SOM). This approach neither did work well – the data were merely randomly distributed and randomly clustered. We were unable to find parameters of the network that worked correctly. In addition, this method was time and computationally consuming.

The most successful method has proven to be the ant colony inspired algorithm (basically the Deneubourg model with improvements from Lumer and Faieta).

In Fig. 7 only ant-colony clustering method itself has been used to lay out vertices and edges. An improvement based on text position (literal distance from record start) has been implemented. It is clear that although the algorithm tried to put together similar items, it actually put together inappropriate words. This is, of course, caused by the measure used (Levenshtein metrics in our case).

4.3   Combined Approach

Although the automated methods provided quite usable results, the human operators were never satisfied and needed to make a correction to the layout as the automated methods lacked the domain knowledge.

An aid of a human expert has been used in semi-automated approach (see Fig. 8 where the automated layout has been corrected by the expert. The correction time has been reduced do about 20–30 seconds only. This approach has proven to be the most effective (and is sometimes referred to as a *doctor-in-the-loop* [8]. After this step we have a good basis for designing the rules for mining.

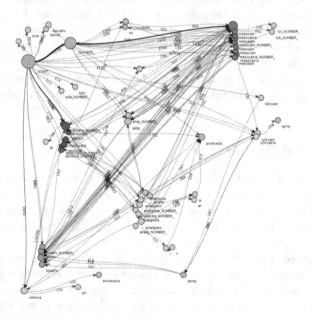

Fig. 1. A human organized transition graph showing the most important relations in one textual attribute. The number literal has been concatenated with the prefix.

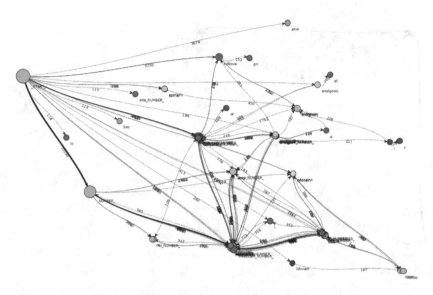

Fig. 2. A human organized transition graph showing the most important relations in one textual attribute. The number literal has been concatenated with the prefix. In this case the expert decided to overlap the nodes.

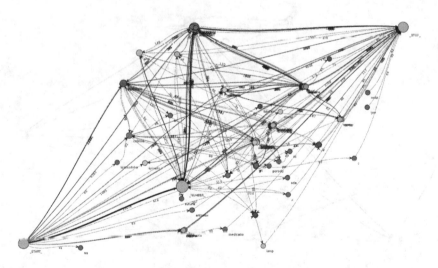

Fig. 3. A human organized transition graph showing the most important relations in one textual attribute. In this case the number literal is taken as a separate node. Moreover, the number of displayed edges has been reduced to improve visibility of other nodes.

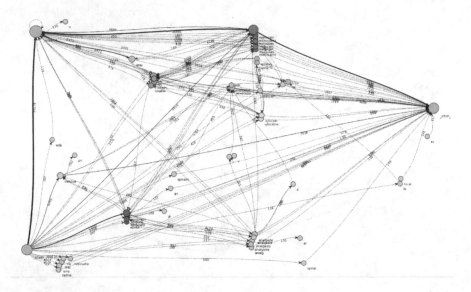

Fig. 4. A human organized transition graph showing the most important relations in one textual attribute. In this case the number literal is taken as a separate node. The expert decided to place the nodes so that they do not overlap.

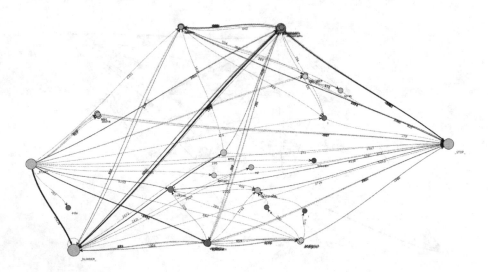

Fig. 5. A human organized transition graph showing the most important relations in one textual attribute. In this case the number literal is taken as a separate literal. The amount of edges displayed has been increased to display more nodes (less frequent). In this case the expert decided to overlap the nodes.

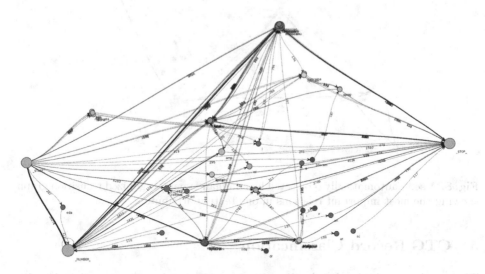

Fig. 6. A human organized transition graph showing the most important relations in one textual attribute. In this case the number literal is taken as a separate literal. In this case the expert decided to overlap the nodes.

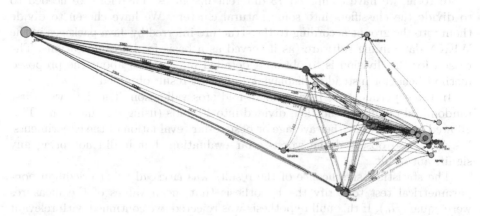

Fig. 7. An sautomatically (not corrected by a human expert) organized transition graph showing the most important relations in one textual attribute.

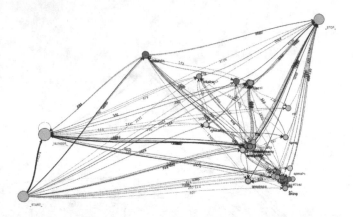

Fig. 8. A semi-automatically (corrected by a human expert) organized transition graph showing the most important relations in one textual attribute.

5   CTG Record Classification

This section describes the classification process that was performed to classify cardiotocography (CTG) records into relevant categories.

In the experimental process we have obtained many measures available for evaluation. In this paper we have evaluated the performance using the *precision* and *recall* measure. We can use (1) the average value (over crossvalidation runs) or (2) maximal (BSF, best-so-far) value. Of course, we have evaluated more measures, such as sensitivity, specificity, f-measure, accuracy, AuROC, auPRC, etc.). However these did not show any significant results.

In total, we have compared 73 different instances. Therefore we needed to to divide the classifiers into some natural clusters. We have chosen to divide them into the groups according to the structure inferred by Java packages in the WEKA datamining software (as it served as a basis for implementation). The reason for the division is to obtain a detailed information about the proposed method behavior instead of displaying only best result obtained.

In the experiment we have used 10-fold crossvalidation. The data were first randomized (shuffled) and then divided into 10-folds (using stratification). This allowed us to obtain either average or best-so-far (evaluation of the experiments). We have also observed the worst-so-far evaluation, but it did not bring any significant result.

The statistical significance of the results was assessed using Friedman nonparametrical test to verify the hypothesis that mean values of the measures were equal (H_0). If this null hypothesis was rejected, we continued with relevant post-hoc tests – Nemenyi test. If a statistically significant difference was found, a Holm and Hochberg post-hoc tests were carried out.

6   Results and Conclusion

In this work we have developed (1) a methodology for effective mining information from large text attributes in a hospital information system and (2) a tool for extracting and visualization of the most prominent parts of large text attributes that did not have any inherent nor explicit structure.

Based on the observation we have implemented basic methods to automatically place the graph nodes: (1) based on the position in the sentence, and (2) using clustering methods. The ant colony inspired clustering has been the most useful for automated grouping related data. Apart from using Euclidean or Cosine similarity, we have used the Levenshtein distance that is advantageous in case of spelling errors and simple mistakes. Short words (up to 4 characters) are still a problem – as the clustering algorithm usually groups them together automatically. A penalty might be introduced if this is a problem, or a certain rules to avoid the problem might be used (such as replacement with constant literals).

The main advantage of the nature inspired concepts lies in automatic finding relevant literals where the group of literals can be adopted by the human analysts and furthermore improved and stated more precisely for use in regular expression design.

The use of induced probabilistic models in the methods increased the speed of analyzing the loosely structured textual attribute analysis and allowed the human analysts to develop lexical analysis grammar more efficiently in comparison to classical methods (the speedup was from about 5–10 min to approx. 20–30 s for an attribute).

We have to note that the expert aid and minor correction is still inevitable. The results of the work are adopted for rule discovery and are designed to be used in expert recommendation system.

In the experimental (classification) phase we have evaluated the average (mean) and BSF measures from 10-fold crossvalidation runs. Also, for each experiment saved and statistically evaluated about 20 quantitative objective measures. The averaged and best-so-far values of the selected measures have been statistically evaluated using Friedman test with Holm and Hochberg post-hoc procedures ($\alpha = 0.05$). The Nemenyi statistical test has also been conducted. Based on the statistical evaluation using the aforementioned methodology, we can conclude that the ACO_DTree methods performed significantly better on the level of $\alpha = 0.05$ when compared to 41 distinct instances of classifiers.

We have experimentally evaluated multiple classifiers over various biomedical data using an WEKA-based implemented framework.

The experimental framework has been implemented with the aim of reproducibility, robustness and statistical soundness of the results. We have also identified the areas various algorithms performed significantly better when compared to other state-of-the art classifier implementations. In addition to the accuracy measure we have evaluated also other quantitative measures.

More than 12 thousand experiments has been run in the final evaluation phase.

7 Discussion and Future Work

The work used CTG data and textual data from a Czech hospital[3], therefore all the text-extraction work was performed in Czech language. This language has not strict rules (for word positioning) when compared for example to the English language. Therefore the records contain many ambiguities, for example it is not always clear to which word the negation (the word *not*) belongs, as it can be positioned before or after the related term.

Also, the medical doctors are used to utilize various non-standard abbreviations and constructs. They also rely on the human processing, so that omitting and "crypting" some information would be "decrypted" by the human (such as typos, omissions of decimal point, etc.). For example, the decimal point in temperature has been often forgotten (resulting in ten times higher patient temperature of 367 °C), using two first letters for plural (*a* for arteria, *aa* for arteriae), using old-habits from typewriting machines (using the letter *o* for zero, *l* for one, so that the string *lo* might be either abbreviation for 'local' or number 10).

Another possible extension of this work would be to search the terms in medical database(s) (literature) and use them to match important keywords. The problem is that no such corpus is available in Czech language: The only corpus available is the Medline translation. As the only way to obtain a directly processable (and validated) inputs is to create a parallel system that allows entering the data in the form we need. This is currently an approach taken in multiple hospitals as the (old) hospital information systems do not allow efficient import.

Acknowledgment. The research is supported by the project No. 15-31398A Features of Electromechanical Dyssynchrony that Predict Effect of Cardiac Resynchronization Therapy of the Agency for Health Care Research of the Czech Republic. This work has been developed in the BEAT research group https://www.ciirc.cvut.cz/research/beat with the support of University Hospital in Brno http://www.fnbrno.cz/en/.

References

1. Adami, C.: Introduction to Artificial Life. Springer Verlag, New York (1998)
2. Blum, C.: Ant colony optimization: Introduction and recent trends. Phys. Life Rev. **2**(4), 353–373 (2005)
3. Burša, M., Lhotská, L.: Ant-inspired algorithms for decision treeinduction - an evaluation on biomedical signals. In: Information Technologyin Bio- and Medical Informatics - 6th International Conference, ITBAM 2015,Valencia, Spain, September 3-4, 2015, Proceedings, pp. 95–106 (2015). http://dx.doi.org/10.1007/978-3-319-22741-2_9
4. Chudacek, V., Spilka, J., Bursa, M., Janku, P., Hruban, L., Huptych, M., Lhotska, L.: Open access intrapartum ctg database. BMC Pregnancy Childbirth **14**, 16 (2014)

---

[3] University Hospital in Brno.

5. Chudáček, V., Spilka, J., Huptych, M., Lhotská, L.: Linear and non-linear features for intrapartum cardiotocography evaluation. Computing in Cardiology 2010 Preprints. New Jersey: IEEE (2015)
6. Dorigo, M., Stutzle, T.: Ant Colony Optimization. MIT Press, Cambridge, MA (2004)
7. Freitag, D., McCallum, A.K.: Information extraction with hmms and shrinkage. In: Proceedings of the AAAI Workshop on Machine Learining for Information Extraction (1999)
8. Holzinger, A.: Interactive machine learning for health informatics: When do we need the human-in-the-loop? Springer Brain Inf. (BRIN) 3(2), 119–131 (2016)
9. Hruban, L., Spilka, J., Chudáček, V., Janků, P., Huptych, M.,Burša, M., Hudec, A., Kacerovský, M., Koucký, M.,Procházka, M., Korečko, V., Segeťa, J., Šimetka, O.,Mchurová, A., Lhotská, L.: Agreement on intrapartumcardiotocogram recordings between expert obstetricians. J Eval Clin Pract, May 2015. http://dx.doi.org/10.1111/jep.12368
10. Lafferty, J., McCallum, A., Pereira, F.: Conditional random fields: Probabilistic models for segmenting and labeling sequence data. In: Proceedings of the ICML, pp. 282–289 (2001)
11. Spilka, J., Chudáček, V., Janků, P., Hruban, L., Burša, M., Huptych, M., Zach, L., Lhotská, L.: Analysis of obstetricians'decision making on CTG recordings. J. Biomed. Inf. 51, 72–79 (2014). http://www.sciencedirect.com/science/article/pii/S1532046414000951

Tumor Growth Simulation Profiling

Claire Jean-Quartier[✉], Fleur Jeanquartier, David Cemernek,
and Andreas Holzinger

Holzinger Group, Research Unit HCI-KDD, Institute for Medical Informatics,
Statistics and Documentation, Medical Informatics, Statistics and Documentation,
Medical University of Graz, Graz, Austria
c.jeanquartier@hci-kdd.org

Abstract. Cancer constitutes a condition and is referred to a group of numerous different diseases, that are characterized by uncontrolled cell growth. Tumors, in the broader sense, are described by abnormal cell growth and are not exclusively cancerous. The molecular basis involves a process of multiple steps and underlying signaling pathways, building up a complex biological framework. Cancer research is based on both disciplines of quantitative and life sciences which can be connected through Bioinformatics and Systems Biology. Our study aims to provide an enhanced computational model on tumor growth towards a comprehensive simulation of miscellaneous types of neoplasms. We create model profiles by considering data from selected types of tumors. Growth parameters are evaluated for integration and compared to the different disease examples.

Herein, we describe an extension to the recently presented visualization tool for tumor growth. The integration of profiles offers exemplary simulations on different types of tumors. The enhanced biocomputational simulation provides an approach to predicting tumor growth towards personalized medicine.

Keywords: Computational biology · Cancer types · Tumor growth · Simulation · HCI · Visualization · Systems biology · Kinetics · Data visualization

1 Introduction

Tumors or neoplasms constitute an overgrowth of tissue, often called mass and lump. Tumors can be non-cancerous (benign) or become cancerous (malignant).

The biological background involves a multistep development based on genomic instability, metabolic reprogramming and evasion from the immune system [1,2]. Underlying signaling events comprise proliferation and growth promotion, cell death inhibition and replicative immortality, neoangiogenesis and invasion [3].

Classification and prediction of the different types of cancer is crucial for therapy and treatment strategies against the often deadly medical condition. Over

© Springer International Publishing Switzerland 2016
M.E. Renda et al. (Eds.): ITBAM 2016, LNCS 9832, pp. 208–213, 2016.
DOI: 10.1007/978-3-319-43949-5_16

the last years, diagnostic classification techniques have made progress advancing from microscopic examination of morphological tissue changes towards gene analysis and biomarker discoveries [4,5]. Respective analysis methods are known for the necessity of computerized support including statistical methods and machine learning classifiers [6]. Yet, methods for predicting tumor growth and cancer prognosis have not fully advanced from basic research to clinical investigations [7]. Still, mathematical and computational models for cancer prediction function as beneficial resource in cancer research. There are several models each on individual types of cancer including brain, ovarian, colon cancers, melanoma, leukemia or head/neck tumors [9]. These models range from deterministic to stochastic, from continuous population dynamics to agent-based individual cell models, from fluid dynamics to Monte Carlo simulations or energy minimization models and are differentially used to describe the various phases of initiation, growth, invasion and migration [10,11]. Moreover, so-called hybrid models simulate variable discrete and continuous aspects on intracellular and intercellular processes [12].

Enhanced modeling tools can provide indications and quantitative criteria for the prognosis of tumor progression as well as therapeutic strategies and support decision making [7,8].

2   Approach

Our study is based on the recently presented simulation on tumor growth [13], which is available at Github: https://github.com/davcem/cpm-cytoscape and as Online Demo: http://styx.cgv.tugraz.at:8080/cpm-cytoscape/. Figure 1 shows a view from the web-tool for tumor growth modeling and its graphical output. The tool depicts an biocomputational approach to simulate biological cell sorting using a two-dimensional extended Potts model [14,15] for the use case of tumor growth on the cellular level.

By scanning through literature and web resources, we consider data from selected types of tumors. These are scanned for usable growth parameters. We create model profiles of exemplary pre-settings for an integration into the simulation. The profiles are compared and evaluated to the different disease examples.

In regard to ease of use for medical scientists, we further improve the graphical user interface for choosing between the model profiles.

3   Results & Discussion

We choose three distinct types of cancer and integrated predefined settings based on recently presented studies on ovarian cancer [16], colorectal cancer [17] and brain cancer [18].

Figure 3 demonstrates the option for choosing between the model profiles for use cases of various types of cancer (Fig. 2).

Table 1 summarizes the selected parameters for the different profiles. Examples are given as custom build as well as ovarian, colorectal and brain cancer

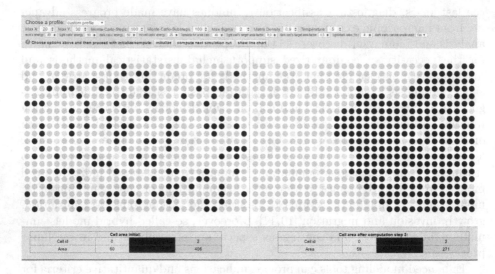

Fig. 1. Screenshot of the tool's graphical view

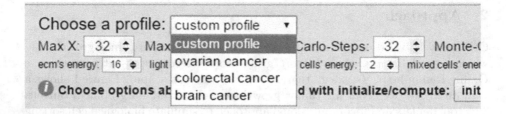

Fig. 2. Screenshot of profile selection via the tool's user interface

Table 1. CPM parameter settings: comparison of settings for custom and profile builds in case of two exemplary cell types corresponding to three different cancer diseases (ovarian, colorectal and brain, adopted from [16–18]).

	Max X $*$ Y	MCS, substeps	Max σ	Matrix density	T
Custom	32 $*$ 32	32, 64	2	0.8	10
Ovarian cancer	162 $*$ 50	10, 135	2	1	10
Colorectal cancer	50 $*$ 110	55, 10	2	0.91	0.1
Brain cancer	50 $*$ 50	250, 10	2	1	4

	J_{ECM}	J_{light}	J_{dark}	J_{mixed}	λ	$A_{t(light)}$	$A_{t(dark)}$	$Ratio_{light/dark}$
Custom	16	15	2	11	0.05	0.4	0.4	1/4
Ovarian cancer	5	-15	-10	-1	3	0.0086	0.0089	1/4
Colorectal cancer	0.2	0.1	0.1	0.2	0.1	0.0046	0.018	1/4
Brain cancer	8	30	3	2	1	1	1	1/2

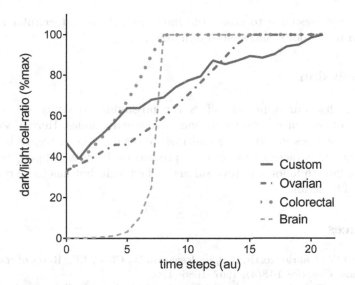

Fig. 3. Cell growth corresponding to various profiles: line chart showing representative ratios between numbers of dark and light cells, given as % max, over computed steps as time.

[13,16–18]. The underlying factors have been adapted in order to be comparable within the tool's framework. While testing identified growth parameters, a key challenge is providing the possibility of dynamic adjustments of cell-related parameters to meet the needs for handling peculiarities of the various use cases of cancer diseases.

The integrated profiles depict distinct characteristics. The three selected examples are compared to the default settings from the custom profile. Representative growth rates differ in kinetic characteristics in terms of relative average and instantaneous growth rate as well as aggressivity and mortality. Figure 3 summarizes ratios between numbers of dark and light cells over time steps, calculated for the different profiles. The use case for brain cancer exhibits the highest growth rate of all selected cases and correlates to exponential growth succeeded by a sudden saturation, which can be translated as the spatial limitation within a biological compartment.

The simplified examples represent basic profiles suitable for likewise human or animal models. Parameters have to be further refined in order to reproduce realistic scenarios for more detailed studies on various aspects of tumor growth, in particular referring to the various cancer diseases.

Due to the variability within cancer classes, refined models are of main interest to supporting the evaluation of tumor aggressiveness and risk assessment.

They provide a resource to study the inter- as well as intracellular system in correlation to temporal and spatial dynamics.

4   Conclusion

We believe that our approach offers the possibility to study tumor growth towards a comprehensive interpretation of tumor dynamics. Herein, we present simulation examples for studying multiple cancer diseases in regard to comparison of kinetic characteristics towards prediction. Results support the notion that tumor growth follows a universal law at first sight but has to be refined and reevaluated in detail.

References

1. Drake, J.W., Charlesworth, B., Charlesworth, D., Crow, J.F.: Rates of spontaneous mutation. Genetics **148**(4), 1667–1686 (1998)
2. Lodish, H., Berk, A., Zipursky, S.L., et al.: Molecular Cell Biology, 4th edn. W. H. Freeman, New York (2000)
3. Hanahan, D., Weinberg, R.A.: Hallmarks of cancer. Cell **674**(5), 646–646 (2011)
4. Cortés, J., et al.: New approach to cancer therapy based on a molecularly defined cancer classification. CA: Cancer J. Clin. **64**(1), 70–74 (2014)
5. Rodríguez-Enríquez, S., Pacheco-Velázquez, S.C., Gallardo-Pérez, J.C., Marn-Hernández, A., Aguilar-Ponce, J.L., Ruiz-García, E., Ruizgodoy-Rivera, L.M., Meneses-García, A., Moreno-Sánchez, R.: Multi-biomarker pattern for tumor identification and prognosis. J. Cell Biochem. **112**(10), 2703–15 (2011)
6. Wang, Y., Tetko, I.V., Hall, M.A., Frank, E., Facius, A., Mayer, K.F., Mewes, H.W.: Gene selection from microarray data for cancer classification-a machine learning approach. Comput. Biol Chem. **29**(1), 37–46 (2005)
7. Vickers, A.J.: Prediction models in cancer care. CA: Cancer J. Clin. **61**(5), 315–326 (2011). doi:10.3322/caac.20118
8. Li, X.L., Oduola, W.O., Qian, L., Dougherty, E.R.: Integrating multiscale modeling with drug effects for cancer treatment. Cancer Inform. **14**(Suppl. 5), 21–31 (2016)
9. Enderling, H., Rejniak, K.A.: Simulating cancer: computational models in oncology. Front Oncol. **3**, 233 (2013)
10. Edelman, L.B., Eddy, J.A., Price, N.D.: In silico models of cancer. Wiley Interdisc. Rev. Syst. Biol. Med. **2**(4), 438–459 (2010)
11. Benzekry, S., Lamont, C., Beheshti, A., Tracz, A., Ebos, J.M., Hlatky, L., Hahnfeldt, P.: Classical mathematical models for description and prediction of experimental tumor growth. PLoS Comput. Biol. **10**(8), e1003800 (2014)
12. Rejniak, K.A., Anderson, A.R.A.: Hybrid models of tumor growth. Wiley Interdiscip Rev. Syst. Biol. Med. **3**(1), 115–125 (2011)
13. Jeanquartier, F., Jean-Quartier, C., Cemernek, D., Holzinger, A.: In Silico Modeling For Tumor Growth Visualization. - Manuscript in revision (2016). https://github.com/davcem/cpm-cytoscape/
14. Graner, F., Glazier, J.A.: Simulation of biological cell sorting using a two-dimensional extended Potts model. Phys. Rev. Lett. **69**, 2013–2016 (1992)
15. Szab, A., Merks, R.M.: Cellular potts modeling of tumor growth, tumor invasion, and tumor evolution. Frontiers in oncology 3 (2013)

16. Giverso, C., Scianna, M., Preziosi, L., Lo Buono, N., Funaro, A.: Individual cell-based model for in-vitro mesothelial invasion of ovarian cancer. Math. Model. Nat. Phenom. **5**(1), 203–223 (2010)
17. Osborne, J.M.: Multiscale model of colorectal cancer using the cellular potts framework. Cancer Inform. **14**(Suppl. 4), 83–93 (2015)
18. Rubenstein, B.M., Kaufman, L.J.: The role of extracellular matrix in glioma invasion: a cellular potts model approach. Biophys J. **95**(12), 5661–5680 (2008)

Integrated DB for Bioinformatics:
A Case Study on Analysis of Functional Effect
of MiRNA SNPs in Cancer

Antonino Fiannaca, Laura La Paglia, Massimo La Rosa[(✉)], Antonio Messina,
Pietro Storniolo, and Alfonso Urso

ICAR-CNR, National Research Council of Italy,
via Ugo La Malfa 153, 90146 Palermo, Italy
{antonino.fiannaca,laura.lapaglia,massimo.larosa,
antonio.messina,pietro.storniolo,alfonso.urso}@icar.cnr.it

Abstract. The era of "big data" arose the need to have computational
tools in support of biological tasks. Many types of bioinformatics tools
have been developed for different biological tasks as target, pathway
and gene set analysis, but integrated resources able to incorporate a
unique web interface, and to manage a biological scenario involving many
different data sources are still lacking. In many bioinformatics approaches
several data processing and evaluation steps are required to reach the
final results. In this work, we face a biological case study by exploiting the
capabilities of an integrated multi-component resources database that
is able to deal with complex biological scenarios. As example of our
problem-solving approach we provide a case study on the analysis of
functional effect of miRNA single nucleotide polymorphisms (SNPs) in
cancer disease.

Keywords: Integrated databases · miRNA SNP · BioGraphDB ·
miRNA-target interactions

1 Introduction

High throughput genomic data coming from biology and translational medicine experiments led scientists to be supported by bioinformatics tools. Different
efforts have been made during these last decades to mine and manage "big data"
coming from next generation sequencing techniques (NGS). This availability of
biological information is still lacking of a complete bioinformatics support that
gives the possibility to store and analyze the obtained data using a common interface and database structure. Indeed, data related to different biological entities
such as proteins, genes, non-coding RNAs, diseases, are generally stored in distinct bioinformatics databases, each one implementing its own data structure
and user interface. Moreover, there is often the necessity to use several of these
resources together due to the complexity of the biological scenarios we want to
investigate. This implies a further time consuming effort in order to skip from

© Springer International Publishing Switzerland 2016
M.E. Renda et al. (Eds.): ITBAM 2016, LNCS 9832, pp. 214–222, 2016.
DOI: 10.1007/978-3-319-43949-5_17

one service to another one, or to transfer data and intermediate results through different resources. In many cases, it is also necessary to deal with synonyms and different accession ids related to the same entity. In this context, the availability of a single framework that can integrate many biological resources and services is a fundamental issue.

Here, we provide a case study that can be solved with different publicly available bioinformatics resources, that is the analysis of functional effect of miRNA single nucleotide polymorphisms (SNPs) in cancer. In order to deal with that case study, we adopted our proposed development framework called BioGraphDB. It is a NoSQL graph database, built upon the OrientDB platform, that collects and links heterogeneous bioinformatics resources. BioGraphDB offers the possibility to face complex bioinformatics scenarios using a single platform, thanks to the integration of different data sources. Graph databases, in fact, with respect to classic SQL systems, allow a greater scalability and queries efficiency with regards to the size of data.

The rest of the paper is organized as follows. In Sect. 2 the most recent integrated databases in bioinformatics will be presented; in Sect. 3 the biological case study will be explained in detail; in Sect. 4 the needed biological resources to face the case study, as well as the BioGraphDB framework adopted in order to solve it, will be described; in Sect. 5 the results obtained with BioGraphDB related to the case study will be presented and discussed; finally in Sect. 6 some conclusions are drawn.

2   Related Work

The large size and different types of biological data have brought the need to design and implement integrated databases that are able to store and to properly link those data. Among the most recent biological databases, mirWalk [4], based on a SQL architecture, integrates data about microRNA (miRNA), messenger RNA (mRNA), miRNA-mRNA interactions, Gene Ontology (GO) annotations [17], pathways and disease. The relationships among these kinds of biological entities can be obtained through a set of pre-defined search methods, that implement precise groups of SQL queries. Typical queries allow to discover gene-miRNA-GO annotations, disease-miRNA relations, gene-miRNA-pathway relations. HumanMine, belonging to the InterMine project [12], is a biological data warehouse system that integrates several biological data including miRNA, genes, proteins, pathways, SNP, diseases and functional associations. Differently from mirWalk, HumanMine offers a query builder service that let the users assemble their own queries in order to obtain the desired information. HumanMine data can also be accessed through web service clients.

More recently, with the development of efficient and flexible NoSQL databases, new ways of integrating and accessing biological data have been investigated [8]. Bio4j [15] is a graphDB, that is a database having as data model a graph structure, focused on the analysis of proteomic data. It in fact integrates data about protein sequences and annotations, GO terms, enzymes. Another graphDB biological database is represented by BioGraphDB [5]. Unlike Bio4j, BioGraphDB offers a wider workbench for bioinformatics studies because it import and integrates in its graph architecture several types of biological data, such as genes and proteins, miRNA and miRNA-target interactions, pathways, GO annotations, sequence data, diseases, scientific literature references. This data integration represents a knowledge base (kb), while the graph organization provides a suitable framework for querying the kb using the most common graph traversal language like Gremlin [1]. BioGraphDB has been previously used for the analysis of different bioinformatics scenarios, like for instance the target analysis of differentially expressed miRNAs in cancer [5]. For those reasons, BioGraphDB is the adopted framework for the proposed case study.

3 Case Study: Analysis of Functional Effect of MiRNA Single Nucleotide Polymorphisms (SNPs) in Cancer

MicroRNAs (miRNAs) are small non coding RNA molecules involved in gene regulation [2]. They act through an imperfect pairing with messenger RNA (mRNA). Different scientific works have showed an important role played by these small molecules in aetiology, progression and prognosis of diseases as cancer [13]. Moreover specific patterns of expression of different miRNAs have been associated to different cancer types [6]. The potential impact of these RNAs in cancer is straightened by the evidence of their mechanism of action: indeed just one miRNA is able to act on different targets, affecting a multitude of genes and leading to widespread phenotypic effects [3]. It is known that variants such as SNPs might affect partially or totally protein functions, altering the normal behavior of that protein in cellular context. To this purpose, Genome-wide association studies (GWAS) aim at identifying genomic regions that contain variants associated with disease. Recently SNPs have been also evidenced in miRNA seeds. Their presence could affect the straightness of interaction with the target, leading to disrupt or create new binding sites [18]. Different works report an association between SNPs and an increased cancer risk [9,20]. Perhaps one of the few examples of a functional association between SNPs and increased risk to develop cancer is described by Jazdzewski [10]. Indeed, they report a polymorphism (rs2910164)in pre-miR-146a that is cause of a lower expression of the mature form and predisposes to papillary thyroid carcinoma. Another interesting work of Saertom [16] showed that the presence of allele variants in miR-125b regulates the expression of BMPR1B gene in an allele-specific way, causing a different colorectal cancer risk depending on the type of allele variant present.

The analysis of miRNA SNPs and the functional impact that allelic variant has on target interaction is important in terms of Gene Ontology analysis because the presence of SNPs could vary the functional impact in a specific cellular context as that of cancer cells.

4   Materials and Methods

4.1   Web Resources for Solving the Proposed Case Study

As previously said the presence of a SNP in a miRNA seed can influence the capability to bound the RNA messenger, thus leading to the creation of new target-binding sites or the disruption of preexisting ones. The analysis of this scenario implies the use of different web resources: miRcancer [19], miRanda [11], miRNASNP [7], and Gene Ontology [17]. In this scenario, starting from a list of differentially expressed (DE) miRNAs in colorectal cancer it is possible to evidence the putative mRNA targets. The first web tool allows to just select over- or down-expressed miRNAs related to a specific cancer disease. Implicitly this web resource is linked to another one, that is miRBase, that contains all the information related to miRNAs, as sequence, structure, mature forms and precursors. Once selected DE miRNAs in colorectal cancer, through resources such as miRNASNP, it is possible to evaluate the impact of the SNP on the miRNA-target interaction. As a consequence of SNP, new binding sites are created and other ones are disrupted. miRNASNP is a web tool containing information on SNPs in miRNA seed, and their related gain or loss of target interactions, allowing to evidence the creation of new interactions. Also for wild-type miRNAs it is necessary to use prediction algorithms to verify what are the putative miRNA targets. One of the tools used to this aim is miRanda. Once identified all the interactions for the two different miRNA lists (miRNA wild-type and miRNA with SNPs), a GO analysis can be performed and all those data can be compared to visualize the changes related to the functional effect of SNPs in miRNA seeds. For this last step of the analysis GO web resources is used. This could be important in better understanding the importance of miRNA role in the pathology of colorectal cancer as they are involved in gene regulation mechanisms. All the discussed steps of the case study have to be done manually by the biologist, with time consuming efforts, using different web interfaces, sometimes managing different kind of data from a resource to another one.

4.2   BioGraphDB Framework

BioGraphDB is a biological integrated graphDB that collects and links together heterogeneous data. It is based on the OrientDB platform [14], which is an open-source NoSQL database management system implemented in Java. The data organization as a graph structure provided by BioGraphDB offers an intuitive and scalable environment for storing and managing of biological data [8]. Biological data modeling as graphs is in fact straightforward. Bio-entities, such as genes or proteins, are mapped into proper instances of vertex (or node) classes. Relationships among those entities are represented as proper instances of edge classes. Moreover, properties of both entities and relationships can be included as attributes of vertexes and edges, respectively. For example, one simple relation such as "the pathway X contains the protein Y" is represented by means of one instance of the *pathway* vertex class and one instance of the *protein* vertex class

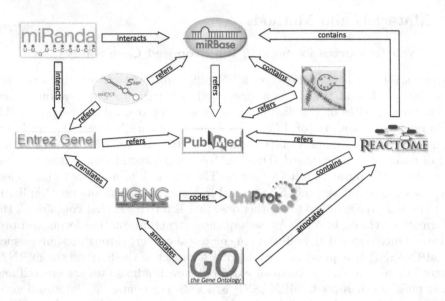

Fig. 1. All the integrated databases in BioGraphDB, with the addition of miRNASNP.

linked by an instance of the *contains* edge class. The mapping process of biological data to their corresponding nodes and edges is done through a set of software modules, written in Java, called Extract-Transformer-Loader (ETL). Currently, BioGraphDB integrates the following databases, where between parenthesis we indicate the type of biological data: NCBI Gene (genes), miRBase (miRNA), miRCancer (miRNA-cancer relations), Pubmed (biomedical literature), UniprotKB (proteins), HGNC (nomenclature), Reactome (pathways), Gene Ontology (functional annotations), miRanda (miRNA-target interactions).

As already explained, BioGraphDB creates a graph composed of all the integrated databases. In this context, each pathway along the graph represents a set of queries. For example, from a starting node, i.e. a miRNA, it is possible to reach a distant node, i.e. a pathway, passing through intermediate linked nodes, i.e. a gene. The identification of this path is equivalent to a query that, given an input miRNA, returns all the pathways involving the target genes of that miRNA. In other words, the graph organization of data allows to solve a very different type and number of bioinformatics scenarios, assuming a path among the involved entities exists. BioGraphDB, thanks to its compliant to the OrientDB platform, can be queried using the high level Gremlin language [1]. Gremlin is a graph traversal language compatible with all the graphDB frameworks implementing the Blueprints property graph data model, like the OrientDB platform. The proposed case study, as described in Sect. 4.1, needs the miRNASNP database that, at moment, is not present in the BioGraphDB framework. For this reason, we developed an ad hoc ETL in order to parse and integrate the miRNASNP data source. All the BioGraphDB integrated databases and their relationships are shown in Fig. 1.

5   Result and Discussion

The proposed graph organization adopted to solve the case study is summarized in Fig. 2. In the center of the figure the BioGraphDB traversal is shown, where circles represent biological entities, and black arrows represent some relationships among these entities. This traversal is divided into tree steps, corresponding to three gremlin queries Q1, Q2 and Q3. Step Q1 (see Fig. 3(a)) allows to select

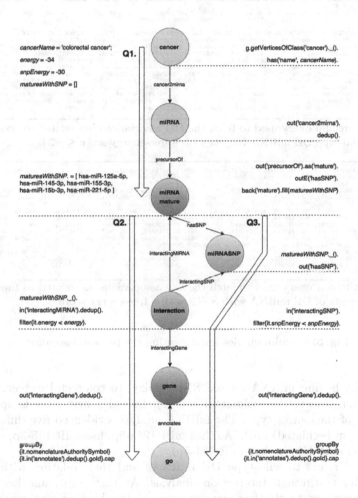

Fig. 2. BioGraphDB traversal for the proposed scenario. The result is given by the combination of three gremlin queries (namely Q1,Q2 and Q3). Circles represent biological entities in BioGraphDB, whereas black arrows represent some relationships among these entities. Each step in graph traversal is reported on the edges of the figure.

```
$ ./gremlin.sh
              \,,,/
              (o o)
-----oOOo-{_}-oOOo-----
gremlin> g = new OrientGraph('remote:localhost/biograph'); cancerName = 'colorectal cancer'; energy = -34; snpEnergy =
-30; maturesWithSNP = []
mag 25, 2016 5:06:10 PM com.orientechnologies.common.log.OLogManager log
INFORMAZIONI: OrientDB auto-config DISKCACHE=13.880MB (heap=455MB os=16.384MB disk=913.218MB)
--->orientgraph[remote:localhost/biograph]

gremlin> g.getVerticesOfClass('cancer')._().has('name', cancerName).out('cancer2mirna').dedup().out('precursorOf').
as('mature').outE('hasSNP').back('mature').fill(maturesWithSNP);maturesWithSNP._().product

==>hsa-miR-125a-5p
==>hsa-miR-145-3p
==>hsa-miR-155-3p
==>hsa-miR-15b-3p
==>hsa-miR-221-5p
```

(a) Q1: Gremlin query used to fetch the DE miRNAs list having miRNA-related SNPs in the colorectal cancer.

```
gremlin> maturesWithSNP._().in('interactingMiRNA').dedup().filter(it.energy < energy).out('interactingGene').dedup().
groupBy{it.nomenclatureAuthoritySymbol}{it.in('annotates').dedup().goId}.cap

==>{TMLHE=[GO:0005506, GO:0005739, GO:0005759, GO:0009083, GO:0016702, GO:0031418, GO:0034641, GO:0044281, GO:0045329,
GO:0050353, GO:0051354, GO:0055114], PSMD13=[GO:0000082, GO:0000165, GO:0000186, GO:0000209, GO:0000278, GO:0000502,
GO:0002223, GO:0002474, GO:0002479, GO:0005198, GO:0005634, GO:0005654, GO:0005829, GO:0005838, GO:0006511, GO:0006521,
GO:0006595, GO:0006915, GO:0006977, GO:0006990, GO:0007127, GO:0007173, GO:0007264, GO:0007411, GO:0008286, GO:0008541,
GO:0008543, GO:0010467, GO:0012501, GO:0016020, GO:0016032, GO:0022624, GO:0031145, GO:0033209, GO:0034641, GO:0038061,
GO:0038095, GO:0042590, GO:0051436, GO:0043066, GO:0043248, GO:0043488, GO:0044281, GO:0045087, GO:0048010, GO:0048011,
GO:0050852, GO:0051436, GO:0051437, GO:0051439, GO:0070062, GO:0090090, GO:0090263], LIF=[GO:0001135, GO:0001974,
GO:0005102, GO:0005125, GO:0005146, GO:0005576, GO:0005515, GO:0005737, GO:0006955, GO:0007275, GO:0007566, GO:0008083,
GO:0008284, GO:0008285, GO:0010976, GO:0016525, GO:0019827, GO:0031100, GO:0031138, GO:0033141, GO:0042503, GO:0042511,
GO:0042577, GO:0043410, GO:0045651, GO:0045835, GO:0045944, GO:0046697, GO:0046888, GO:0048286, GO:0048644, GO:0048656,
GO:0048708, GO:0048711, GO:0048861, GO:0048863, GO:0050731, GO:0051461, GO:0060041, GO:0060290, GO:0060426, GO:0060463,
GO:0060707, GO:0060708, GO:0070373, GO:0072108, GO:0072307, GO:1900182, GO:1901676, GO:1903025]}
```

(b) Q2: Gremlin query used to fetch the GO associations list related to those predicted wild-type DE miRNA targets with a free-energy score ≤ -34.

```
gremlin> maturesWithSNP._().out('hasSNP').in('interactingSNP').filter(it.snpEnergy < snpEnergy).out('interactingGene').
dedup().groupBy{it.nomenclatureAuthoritySymbol}{it.in('annotates').dedup().goId}.cap

==>{PTP4A2=[GO:0004727, GO:0005634, GO:0005737, GO:0005769, GO:0005886, GO:0035335, GO:0070062], SYK=[GO:0001525,
GO:0001945, GO:0002092, GO:0002223, GO:0002250, GO:0002281, GO:0002283, GO:0002366, GO:0002554, GO:0004672, GO:0004674,
GO:0004713, GO:0004715, GO:0004716, GO:0005178, GO:0005515, GO:0005524, GO:0005634, GO:0005737, GO:0005829, GO:0005886,
GO:0006468, GO:0007159, GO:0007169, GO:0007229, GO:0007257, GO:0007596, GO:0008283, GO:0009887, GO:0010543, GO:0010803,
GO:0016032, GO:0018105, GO:0019370, GO:0019815, GO:0019901, GO:0030154, GO:0030168, GO:0030593, GO:0031234, GO:0031623,
GO:0032009, GO:0032928, GO:0033630, GO:0038083, GO:0038095, GO:0038096, GO:0042101, GO:0042742, GO:0042991,
GO:0043234, GO:0043306, GO:0043313, GO:0043366, GO:0045087, GO:0045401, GO:0045425, GO:0045579, GO:0045588, GO:0045780,
GO:0046638, GO:0046641, GO:0048514, GO:0050731, GO:0050764, GO:0050850, GO:0050853, GO:0051090, GO:0070372,
GO:0071226, GO:0071404, GO:0090237, GO:0090330], SMYD5=[GO:0003674, GO:0005575, GO:0008150, GO:0008168, GO:0032259,
GO:0046872], MIR4313=[], LGR6=[GO:0004888, GO:0004930, GO:0005515, GO:0005887, GO:0007186, GO:0016055,
GO:0030177, GO:0030335, GO:0031982, GO:0032588, GO:0090263], TXNL4A=[GO:0000245, GO:0000375, GO:0000398, GO:0005515,
GO:0005654, GO:0005681, GO:0005682, GO:0007067, GO:0008380, GO:0010467, GO:0046540, GO:0051301], CRHR1-
IT1=[GO:0005576]}
```

(c) Q3: Gremlin query used to fetch the GO associations list related to those predicted targets of DE miRNA with SNPs with a free-energy score ≤ -30.

Fig. 3. Gremlin queries implementing the proposed scenario.

DE miRNA having miRNA-related SNPs linked to colorectal cancer, starting from a given disease (i.e. colorectal cancer). Indeed a first filter is applied on the choice of the cancer type. The miRNA analysis evidenced five differentially expressed (up-regulated) miRNAs (hsa-miR-125a-5p, hsa-miR-145-3p, hsa-miR-155-3p, hsa-miR-15b-3p, hsa-miR-221-5p).

Step Q2 refers to wild-type DE miRNAs and their relation with specific mRNA targets through interaction analysis. At this point, another filter is applied according to the free energy score of the binding site predicted by miRanda. This allows to highlight only miRNA-target interactions that are strongly bound. The targets evidenced are then analyzed through GO enrichment, to see the functional annotations that link these molecules to colorectal cancer development. Step Q3 refers to miRNA with SNPs analysis through a parallel way to Q2. Also in this step, a filter on free energy score is applied. Results of step Q2 and Q3 are showed in Fig. 3(b) and (c) respectively.

The two target-interaction analysis showed the binding of TMLHE, PSMD13 and LIF genes, with wild-type miRNAs; the binding of PTP4A2, SYK, SMYD5, MIR4313, LGR6, TXNL4A, and CRHR1-IT1 with miRNAs with SNPs. As previously said, after analyzing the miRNA-target interaction predictions, we wanted to study the gene target functional enrichment of the obtained lists (the obtained results are showed in Fig. 3(b) and (c) respectively for wild-type miR-NAs and miRNAs with SNPs). The GO analysis of the two target lists evidenced many associated GO annotations. The comparison between these GO lists evidenced some terms associated with important functional processes as regulation of cell proliferation, growth factor activity, negative regulation of angiogenesis, regulation of mRNA stability, positive regulation of MAPK cascade, negative regulation of apoptotic process, transcription factor activity, RNA polymerase II transcription factor recruiting. More in detail, the gene group related to the miRNAs with SNPs, was strongly associated with gene expression, immune system and signal transduction, but it was not associated to programmed cell death and cell cycle as the gene group linked to wild-type miRNAs.

6   Conclusions

Thanks to the continuous growing and availability of different types of biological data, it is possible to face complex bioinformatics scenarios using a lot of publicly available web resources. In many cases, however, there is not a unique framework that stores and makes accessible all the needed resources, forcing this way a bioinformatician to use different web tools, dealing with issue related to interoperability and intermediate results to move from one system to another one. In this context, we proposed the use of an integrated biological database, namely BioGraphDB, in order to face a case study based on the analysis of functional effect of miRNA SNPs in cancer. The occurrence of SNPs on miRNA seeds can, in fact, change some of the miRNA targets, adding new target or losing targets related to the miRNA wild type, altering this way the functional impact in a specific cellular context as that of cancer cells. For this case study, BioGraphDB offers a suitable framework for analysis. It integrates all the needed biological resources and by means of its graph architecture allows the user to navigate the graph moving from one resource to another one. Each graph traversal, implemented by means of the Gremlin language, represents a query over the BioGraphDB. In this work, we showed how three queries, together with the corresponding path along BioGraphDB, produced the functional analysis that was the focus of the proposed case study.

References

1. Apache Software Foundation: Apache TinkerPop. http://tinkerpop.incubator. apache.org

2. Bartel, D.P.: Micrornas: target recognition and regulatory functions. Cell **136**(2), 215–233 (2009)

3. Calin, G.A., Sevignani, C., Dumitru, C.D., Hyslop, T., Noch, E., Yendamuri, S., Shimizu, M., Rattan, S., Bullrich, F., Negrini, M., et al.: Human microrna genes are frequently located at fragile sites and genomic regions involved in cancers. Proc. Natl. Acad. Sci. USA **101**(9), 2999–3004 (2004)

4. Dweep, H., Gretz, N., Sticht, C.: miRWalk database for miRNA-target interactions. Methods Mol. Biol. **1182**, 289–305 (2014)

5. Fiannaca, A., La Rosa, M., La Paglia, L., Messina, A., Urso, A.: BioGraphDB: a new GraphDB collecting heterogeneous data for bioinformatics analysis. In: Proceedings of BIOTECHNO 2016. IARIA (in press)

6. Garzon, R., Calin, G.A., Croce, C.M.: Micrornas in cancer. Ann. Rev. Med. **60**, 167–179 (2009)

7. Gong, J., Tong, Y., Zhang, H.M., Wang, K., Hu, T., Shan, G., Sun, J., Guo, A.Y.: Genome-wide identification of SNPs in microRNA genes and the SNP effects on microRNA target binding and biogenesis. Human Mutat. **33**(1), 254–263 (2012)

8. Have, C.T., Jensen, L.J.: Are graph databases ready for bioinformatics? Bioinformatics **29**(24), 3107–3108 (2013)

9. Iwai, N., Naraba, H.: Polymorphisms in human pre-mirnas. Biochem. Biophys. Res. Commun. **331**(4), 1439–1444 (2005)

10. Jazdzewski, K., Murray, E.L., Franssila, K., Jarzab, B., Schoenberg, D.R., de La Chapelle, A.: Common snp in pre-mir-146a decreases mature mir expression and predisposes to papillary thyroid carcinoma. Proc. Natl. Acad. Sci. **105**(20), 7269–7274 (2008)

11. John, B., Enright, A.J., Aravin, A., Tuschl, T., Sander, C., Marks, D.S.: Human microRNA targets. PLoS Biol. **2**(11), e363 (2004)

12. Kalderimis, A., Lyne, R., Butano, D., Contrino, S., Lyne, M., Heimbach, J., Hu, F., Smith, R., Pan, T., Sullivan, J., Micklem, G.: InterMine: extensive web services for modern biology. Nucleic Acids Res. **42**(W1), W468–W472 (2014)

13. Kumar, M.S., Lu, J., Mercer, K.L., Golub, T.R., Jacks, T.: Impaired microrna processing enhances cellular transformation and tumorigenesis. Nature Genet. **39**(5), 673–677 (2007)

14. Orient Technologies LTD: OrientDB. http://orientdb.com

15. Pareja-Tobes, P., Tobes, R., Manrique, M., Pareja, E., Pareja-Tobes, E.: Bio4j: a high-performance cloud-enabled graph-based data platform. Technical report, Era7 bioinformatics (2015). http://biorxiv.org/lookup/doi/10.1101/016758

16. Sætrom, P., Biesinger, J., Li, S.M., Smith, D., Thomas, L.F., Majzoub, K., Rivas, G.E., Alluin, J., Rossi, J.J., Krontiris, T.G., et al.: A risk variant in an mir-125b binding site in bmpr1b is associated with breast cancer pathogenesis. Cancer Res. **69**(18), 7459–7465 (2009)

17. Nucleic Acids Res. Gene Ontology Consortium: going forward. **43**(D1), D1049–D1056 (2015)

18. Thomas, L.F., Saito, T., Sætrom, P.: Inferring causative variants in microrna target sites. Nucleic Acids Res. **39**, 109 (2011). p. gkr414

19. Xie, B., Ding, Q., Han, H., Wu, D.: miRCancer: a microRNA-cancer association database constructed by text mining on literature. Bioinformatics **29**(5), 638–644 (2013)

20. Yang, H., Dinney, C.P., Ye, Y., Zhu, Y., Grossman, H.B., Wu, X.: Evaluation of genetic variants in microrna-related genes and risk of bladder cancer. Cancer Res. **68**(7), 2530–2537 (2008)

The Database-is-the-Service Pattern for Microservice Architectures

Antonio Messina, Riccardo Rizzo[✉], Pietro Storniolo, Mario Tripiciano, and Alfonso Urso

ICAR-CNR, Palermo, Italy
{antonio.messina,riccardo.rizzo,pietro.storniolo,
mario.tripiciano,alfonso.urso}@icar.cnr.it

Abstract. Monolithic applications are the most common development paradigm but they have some drawback related to the maintenance, upgrading and scaling. Microservice architectures were recently proposed in order to solve some of these issues, because they are simpler to scale and more flexible. Both architectures use a database and this component can act as a component for micro service. In the paper we present a pattern for microservice architecture that uses a database as component, and this pattern is used in an health record application. We explain also the requirements of the database for this pattern and the advantages achieved.

Keywords: Microservices · Scalable applications · Continuous delivery · Microservices patterns · noSQL · Database

1 Introduction

Monolithic applications force the developers to work in large teams that tend to build complex applications, difficult to maintain, and understand, these applications can increment the transaction volume only by running in multiple copies that often access to the same database. Recently the paradigm of microservices [7] has been adopted in application development in order to build more flexible applications, simpler to build and to deploy. In microservices based systems one of the biggest challenge is the partition into separated services, each of them should be simple enough to have a small set of responsibilities. This fragmentation of the application can make data management one of the bottleneck of the system, even if one or few micro services are only application that access the data. Microservices can become an effective architectural pattern if these design issues are solved, in fact studies have shown how this architectural pattern can give benefits when enterprise applications are deployed in cloud environments [17] and in containers [2], e.g. Docker [3]. The microservices architectural pattern is also considered the natural fit for the Machine-to-Machine (IoT) development [12]. In the new microservice pattern proposed in this paper, a database, under certain circumstances and thanks to the integration of some business logic, can be considered a microservice by itself. It will be labeled as the database-is-the-service pattern.

© Springer International Publishing Switzerland 2016
M.E. Renda et al. (Eds.): ITBAM 2016, LNCS 9832, pp. 223–233, 2016.
DOI: 10.1007/978-3-319-43949-5_18

2   Background

In the next section we will introduce the microservice architecture style, but, to better understand the underlying philosophy, we will give here a short description of a classical monolithic application. Typical business applications are made up of three distinct entities: a client-side user interface that deals with data visualisation and with the user interaction; server side unit that receives requests from the client, performs the logical domain and retrieves and updates the relational databases that constitutes the third part of the application.

A monolith application has all its functionality grouped into a single unit or application (for example a single JAR, WAR or EAR file). These applications are widely known and can be found in several existing tools such as application servers, frameworks, scripts and so on. Applications developed in that way, even if built as distributed application, already include all the features and services all together and immediately available. Moreover also the testing phase of a monolithic application is easier to perform since it has no dependencies. However, monolithic applications has many disadvantages: before starting the development of a monolithic application the technology to be used must be carefully evaluated. The whole team will have to use the same language and tools to facilitate the alignment of the various working groups. Often it will be impossible to update the application with new technologies or introduce new features because that would need the rewriting of the entire application. The only possible scalability of these type of applications is the duplication on different servers with obvious reduction of cache efficiency, memory usage and I/O traffic. In fact monolith applications can only scale in one dimension, i.e. they have to be entirely duplicated across a set of servers (see Fig. 1). This way, each application instance will access all of the data. Systems based on microservices present many advantages. Some of them came from their distributed architecture and will be explained in the next section.

3   The Patterns Related to the Microservice Architectures

The microservices architectural pattern requires the functional decomposition of an application. The application to implement needs to be broken down into multiple smaller services, each deployed in its own archive, and then composed as a single application using standard lightweight communication, such as REST over HTTP (see Fig. 2).

Basically, the service design should be made by applying the Single Responsibility Principle [8], that defines a responsibility of a class as a reason to change, and states that a class should only have one reason to change. There are several patterns [16] related to the microservices pattern. We mainly focus our attention on the following:

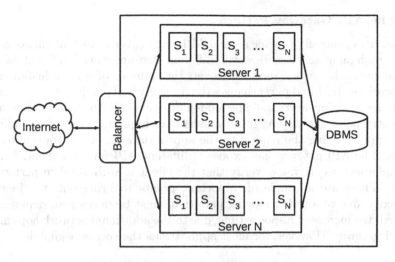

Fig. 1. Service provided by horizontal scaled monolithic application

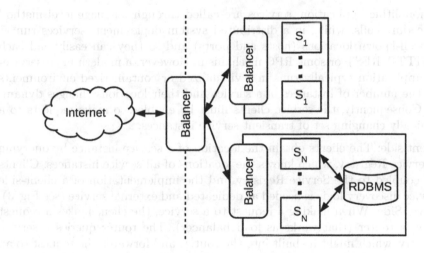

Fig. 2. Tipical microservice-based application with lightweight frontend

- The API Gateway pattern, that defines how clients access the services in a microservices architecture.
- The Client-side and Server-side Discovery patterns, used to route requests for a client to an available service instance in a microservices architecture.
- The Service Registry pattern, a critical component that tracks the instances and the locations of the services.
- The Database per Service pattern, that describes how each service has its own database.

3.1   The API Gateway Pattern

Microservices generally provide several APIs, and this means that clients need to interact with many services. However, different customers need different data and different network performance. Moreover, the number of service instances and their locations (host and port) changes dynamically and partitioning into services can change over time and should be hidden from clients. All clients should have a single entry point gateway API; this one handles requests in one of two following ways: simply sending the request to the appropriate service or routing to more services. The API gateway can expose a different API for each client. It might also implement security, e.g. verify that the client is authorized to perform the request. There are a couple of important points to bring out: the increased complexity, due to another moving part that must be developed, deployed and managed; the increased response time, due to the additional network hop through the API gateway. However, for most applications the cost is negligible.

3.2   The Discovery Patterns

In a monolithic application, services are called through language-level method or procedure calls, while, in a distributed system deployment, services run at fixed, well-known locations (hosts and ports) and so they can easily call each using HTTP/REST or some RPC mechanism. However, a modern microservice-based application typically runs in a virtualized or containerized environments where the number of instances of a service and their locations changes dynamically. Consequently, the service clients must be enabled to make requests to a dynamically changing set of transient service instances.

- Client-side: The clients obtain the location of a service instance by querying a Service Registry, which knows the locations of all service instances. Clients are coupled to the Service Registry and the implementation of a client-side service discovery logic is needed as dedicated and external service (see Fig. 3).
- Server-Side: When making a request to a service, the client makes a request using a router (that works as load balancer). The router queries a service registry, which might be built into the router, and forwards the request to an available service instance. Compared to client-side discovery explained before, the client code is simpler since it does not have to deal with discovery, but more network hops are required (see Fig. 4).

3.3   The Service Registry Pattern

A service registry is a database of services, constituted by their instances and their locations. The service instances are registered with the service registry on start-up and deregistered on shutdown. Client of the service and/or routers ask the service registry to find the available instances. Usually this one is a different infrastructure component that needs the setup and configuration. In addition, the Service Registry is a critical component of the system and must be highly available.

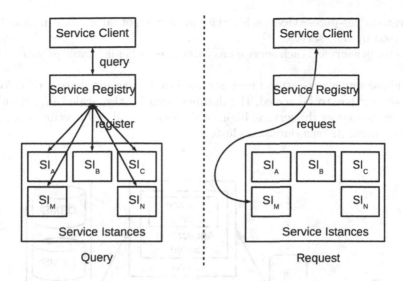

Fig. 3. Client-side discovery pattern

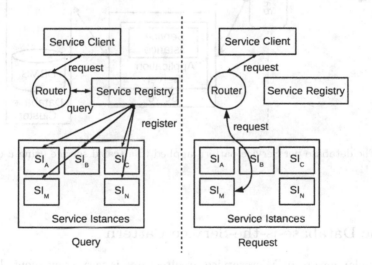

Fig. 4. Server-side discovery pattern

3.4   The Database per Service Pattern

According to this pattern, we should keep all the persistent data of a micro service private to that service and accessible only via its API. It means that the database of the service is effectively part of the implementation of that service and not directly accessible by other services. There are a few different ways to keep a service's persistent data private:

- Private-tables-per-service: each service owns a set of tables that must only be accessed by that service.
- Schema-per-service: each service has a database schema that is private to that service
- Database-server-per-service: each service has its own database server. When the service has to be scaled, the database can be also scaled in a database cluster, no matter the service. Figure 5 shows a typical architecture of a scaled service using its own database cluster.

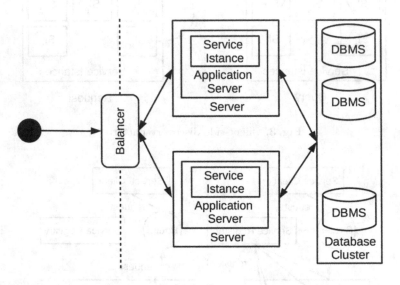

Fig. 5. The database per Service pattern applied to a scaled service using a database cluster

4   The Database-is-the-Service Pattern

The granular nature of Microservice architectures is very convenient, but also has a greater complexity. Split a monolith in microservices makes each component easier to develop, but it becomes very complex to put all the components together and to integrate the various parts [5]. So there is a kind of complexity conservation law about the development of software architectures [4]. Moreover, the sophisticated demands of today's businesses make the complexity growing at a dizzying rate and this increases the risks of potential blacks holes and loss of data. Obviously Microservice architectures can amplify these risks, because of their distributed nature and the amount of calls to remote procedures. Moreover the developers have to work on many factors such as network latency, fault tolerance, version control, variable loads etc. A possible approach to reduce complexity and to gain in speed and scalability may be the addition of new behaviours and business logic at the database level starting from the Database-Server per Service Pattern. Assuming that each scalable service has its own database, in order to reduce the architectural complexity and the risk that originates from this complexity we decided to embed in the database the business logic that implements the desired service. In this way the service is strictly coupled to the data, hence this pattern is even stronger than the Database-Server per Service Pattern, because the database itself acts as a business service. In Fig. 6 you can see how the clients requests are routed via a load balancer, following the guidelines of the Server-side Discovery Pattern.

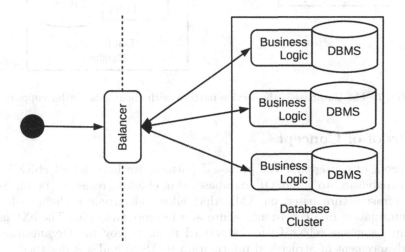

Fig. 6. The database is the service pattern: DBMS with business logic

The proposed approach has some clear advantages: (a) the traditional service layer disappears, thanks to the whole removal of related hosts and application servers or containers; (b) services deployed into the database have instant access to data; (c) a smaller number of components with less complexity, and therefore less risk. Unlike Client-side Discovery Pattern, there's no need to implement a discovery logic into clients, there isn't any balancer, and the cluster layer supplies, at least, the same Service Registry capabilities. Figure 7 shows a representation of the architecture. Drawbacks are also obvious, first and foremost the dependency on the chosen database, because the service becomes integral to, and inseparable from, the database engine. Test and debug activities must also involve the database because of his primary role, and for this reason, the database source code must be available.

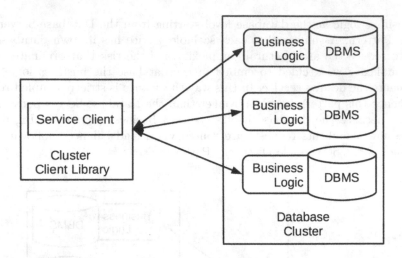

Fig. 7. The Database is the service pattern with client side cluster support

5   Proof of Concept

As a proof of concept for the proposed pattern, we have added ebXML registry capabilities to a noSQL database. An ebXML registry [14,15] is an
open infrastructure based on XML that allows electronic exchange of business information in a consistent, secure and interoperable way. The eXtensible
Markup Language (ebXML) is a standard promoted by the Organisation for
the Advancement of Structured Information (OASIS) and was designed to create a global electronic marketplace where enterprises of any size, anywhere, can
find each other electronically and conduct business using exchange of XML messages. Nowadays, ebXML concepts and specifications are reused by the Cross-
Enterprise Document Sharing (XDS) architectural model [13], defined by Integrating the Healthcare Enterprise (IHE) initiative [1], which is promoted by
healthcare professionals and industries to improve the way computer systems in
healthcare share information. We developed this pattern starting from EXPO
[11], a prototypal extension for OrientDB derived from eRIC [9], our previous
SQL-based ebXML registry implemented as web service in Java. In the following subsections we introduce the noSQL database engine we have chosen and we
briefly illustrate what we have done and some interesting performances results.

5.1   A Multi-model NoSQL DBMS as an EbXML Registry

OrientDB is also a customisable platform to build powerful server component
and applications, it also contains an integrated web server, so that it is possible to create server side applications. The customisations can be obtained by
developing new handlers, to build plugins that start when OrientDB starts, or
implementing custom commands, the suggested best way to add custom behaviours or business logic at the server side. The OrientDB Object Interface works on

top of the Document-Database and works like an Object Database by managing directly Java objects. This makes things easier for the Java developer, since the binding between objects to records is transparent. These objects are referred as POJOs: Plain Old Java Objects. The ebXML objects are perfect POJOs, because they are serialisable, have a no-argument constructor, and allow access to properties using getter and setter methods that follow a simple naming convention. They are also fully described by the standard set of Java XML annotation tags because they need to be transferred over the line properly encapsulated using the Java Architecture for XML Binding (JAXB) [6]. This means that we can add new custom properties preceded by the @XMLTransient tag without breaking things. We have used those new properties and the related getter and setter methods to add native OrientDB links between objects, which can be transparently serialized/deserialized by the OrientDB engine in their enriched form. This approach has a great impact on the management of ebXML objects: (a) they are still used in the standard way within the client-server SOAP interactions; (b) the binding to the database records is transparent; (c) there is no need of extra data class object (DAO) hierarchy to manage the persistence; (d) we are able to make full use of the OrientDB capabilities. An extension of the Orient-dDB OServerCommandAbstract class has replaced the old Java servlet in the registry requests management. In particular, the execute() method is invoked at every HTTP request and let us to read the input HTTP stream and to write into the output HTTP stream. This is the place where we intercept, elaborate

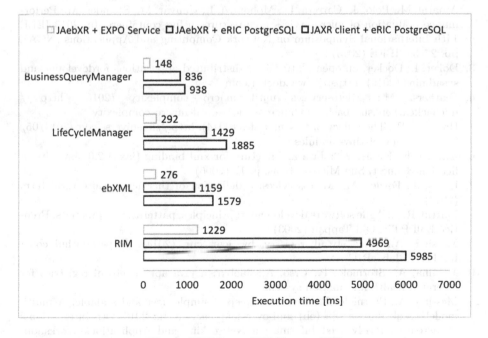

Fig. 8. EXPO vs eRIC/PostgreSQL

and reply to the incoming requests, by calling the real ebXML registry layer. To verify ebXML specifications compliance and to evaluate the performances, we have run the same ebXML test suite developed with the JAebXR API [10] and then we have compared the results in some different configurations. Figure 8 shows the interesting performance improvements obtained with EXPO service.

6   Conclusions and Future Work

While monoliths have been the norm for some time, microservices have emerged as an alternative to deal with certain limitation in monoliths. However, that doesn't mean that monoliths are completely obsolete. Obviously, it's important to look at advantages and disadvantages of each and, as much information as possible, to make the aware decision. In this paper we introduced a new microservice pattern where the database is the service. The proposed pattern has been tested adding ebXML registry capabilities to a noSQL database, experimental tests show improved performances of the proposed simplified microservice architecture compared with SQL-based ebXML registry implemented as traditional Java web service.

References

1. Integrating the healthcare enterprise (ihe) (2010). http://ihe.net
2. Amaral, M., Polo, J., Carrera, D., Mohomed, I., Unuvar, M., Steinder, M.: Performance evaluation of microservices architectures using containers. In: 2015 IEEE 14th International Symposium on Network Computing and Applications (NCA), pp. 27–34. IEEE (2015)
3. Doker, I.: Docker, an open platform for distributed applications fordevelopersand sysadmins (2016). https://www.docker.com
4. Feathers, M.: Microservices until macro complexity (2016). https://michaelfeathers.silvrback.com/microservices-until-macro-complexity
5. Hammer, H.: The fallacy of tiny modules (2014). http://hueniverse.com/2014/05/30/the-fallacy-of-tiny-modules
6. Kawaguchi, K.: Jsr 222: Java architecture for xml binding (jaxb) 2.0. Java Specification Request, Sun Microsystems, p. 17 (2006)
7. Lewis, J., Fowler, M.: Microservices: a definition of this new architectural term (2014)
8. Martin, R.C.: Agile software development: principles, patterns, and practices. Prentice Hall PTR, Old Tappan (2003)
9. Messina, A.: eric: ebxml registry by icar cnr (2014). https://github.com/IcarPA-TBlab/eRIC
10. Messina, A., Storniolo, P., Urso, A.: Jaebxr: a java api for ebxml registries for federated health information systems
11. Messina, A., Storniolo, P., Urso, A.: Keep it simple, fast and scalable: A multi-model nosql dbms as an (eb) xml-over-soap service. In: 2016 30th International Conference on Advanced Information Networking and Applications Workshops (WAINA), pp. 220–225. IEEE (2016)

12. Namiot, D., Sneps-Sneppe, M.: On micro-services architecture. Int. J. Open Inf. Technol. **2**(9), 24–27 (2014)
13. Noumeir, R.: Sharing medical records: The xds architecture and communication infrastructure. IT Prof. Mag. **13**(4), 46 (2011)
14. OASIS, ebXML Registry Technical Committee: Registry information model (rim) v3.0 (2005). http://docs.oasis-open.org/regrep/regreprim/v3.0/regrep-rim-3.0-os.pdf
15. OASIS, ebXML Registry Technical Committee: Registry services and protocols v3.0(2005). http://docs.oasis-open.org/regrep/regreprs/v3.0/regrep-rs-3.0-os.pdf
16. Richardson, C.: Microservice architecture patterns and best practices. http://microservices.io/index.html
17. Villamizar, M., Garces, O., Castro, H., Verano, M., Salamanca, L., Casallas, R., Gil, S.: Evaluating the monolithic and the microservice architecture pattern to deploy web applications in the cloud. In: Computing Colombian Conference (10CCC), 2015 10th, pp. 583–590. IEEE (2015)

A Comparison Between Classification Algorithms for Postmenopausal Osteoporosis Prediction in Tunisian Population

Naoual Guannoni[1(✉)], Rim Sassi[2], Walid Bedhiafi[2],
and Mourad Elloumi[1]

[1] Laboratory of Technologies of Information and Communication and Electrical Engineering (LaTICE), National Superior School of Engineers of Tunis (ENSIT), University of Tunis, Tunis, Tunisia
nawel.gannouni90@gmail.com, mourad.elloumi@gmail.com
[2] LR05ES05 Laboratory of Genetic,
Immunology and Human Pathologies (LGIPH),
Faculty of Sciences of Tunis,
Université de Tunis El Manar, 2092 Tunis, Tunisia
sasrim2006@yahoo.com, walidbedhiafi@hotmail.com

Abstract. In this paper, we make an experimental study to compare the performances of different data mining classification algorithms for predicting osteoporosis in Tunisian postmenopausal women. This study aims to identify the best algorithms with the optimum classification parameters values and to determine the most important risk factors that have a significant impact on the osteoporosis occurrence. The obtained results show that *Support Vector Machine* (SVM) classifier and *Artificial Neural Network* (ANN) classifier give the best classification performances when dealing with the three bone statuses (*normal, osteopenia, osteoporosis*). On the other hand, the *decision tree* classifier C4.5 enables to extract the most important risk factors for osteoporosis occurrence. The selected risk factors are validated by biologists.

Keywords: Data mining · Classification algorithms comparison · Optimum parameters · Prediction · Osteoporosis risk factors

1 Introduction

According to the World Health Organization (WHO) report, "osteoporosis is a skeletal disorder characterized by low bone mass and micro architectural deterioration of bone tissue, resulting in skeletal fragility and an increased risk of fracture". This disease is characterized by reduced *Bone Mineral Density* (BMD) and change in bone proteins concentrations [1, 2].

Osteoporosis is common in postmenopausal women because the bone mass decreases with age and with the drastic hormonal changes (decrease in estrogen level) [7].

This complex disease becomes more and more a worldwide major public health issue. Since today, it is estimated that 200 million people are suffering from this d ease

© Springer International Publishing Switzerland 2016
M.E. Renda et al. (Eds.): ITBAM 2016, LNCS 9832, pp. 234–248, 2016.
DOI: 10.1007/978-3-319-43949-5_19

in the world and that 30 % menopausal women in Europe and the United States have osteoporosis [3, 4].

Osteoporosis modeling and prediction is an emerging knowledge-based paradigm that exploits the power of computational methods in order to make the best decision and to optimize the efficiency of disease treatment [5]. Indeed, the demand for patient's management with this disease will increase. This is associated with innovative research for modeling methods. Early detection of osteoporosis is not an easy process, but if this disease is detected, it is curable.

In bibliography, a lot of studies have been undertaken to predict osteoporosis using classification models for different populations. In [6], Chang et al. compared computational tools with and without feature selection for predicting osteoporosis outcome in a Taiwanese women population. Three classification algorithms were applied in this study: *Multilayer Neural Network*, *Naive Bayes* and *Logistic Regression*. In [7], Yoo et al. developed the prediction models to accurately identify the risk of femoral neck fracture in postmenopausal Korean women using various machine learning methods such as *Support Vector Machine* (SVM), *Random Forest* (RF), *Artificial Neural Network* (ANN) and *Logistic Regression* (LR). In [1], Iliou et al. evaluated twenty machine learning techniques to categorize Greek patients into osteoporosis and non osteoporosis classes. In the study of Xu et al. [8], several combined features were used to construct a classifier for predicting osteoporosis. SVM and KNN techniques are introduced to build classifiers with these features.

However, there is no study that uses classification algorithms for postmenopausal osteoporosis detection in Tunisian population. Since osteoporosis is a multifactor disease characterize by specific genetic and environmental factors, combination of risk factors may vary from population to other. Thus, models and methods for risk prediction should be adapted to each population. Indeed, risk factors seem to be different and specific for Tunisian population. For this reason, we use in this paper data mining techniques to better classify patients with normal *Bone mineral density* (BMD), with *osteopenia* or *osteoporotic* in a Tunisian sample. In addition, we find the effectiveness of each classification algorithms and we identify the best classifier with the optimal parameters. After that, we extract the most important risk factors for osteoporosis occurrence using the decision tree algorithm. The experimental study is analyzed using Weka system. The best classifiers found will be improved and used to better predict if a women can remain with osteopenia or will becomes osteoporotic in order to make the best decision.

This paper is structured as follows: Sect. 2 presents a review of classification algorithm used in this study. Section 3 follows reporting the experimental study and the practical application of our cohort for classification algorithms. In Sect. 4 we discuss our obtained results and we compare them with results of biologists. Finally, in Sect. 5, we conclude by summarizing our results, and sketching directions of further research.

2   Classification Techniques: A Review of Classification Algorithms

Our methodology consists to apply each classifier on a data set of postmenopausal osteoporosis in Tunisian population to determine the best classification algorithm and to extract the most important risk factors for osteoporosis occurrence.

We choose a broad set of six different classifiers to find the effectiveness of each algorithm for predicting the state of the bone as the following:

1. Decision tree: is a very popular classification technique where each internal node denotes a test on the value of one or more attributes, each branch represents an outcome of the test which corresponds to attribute values, and each terminal node holds a class label. In this paper, we choose C4.5 for performance analysis which is known as j48 in Weka [9].
2. Artificial Neural Network (ANN): is a set of input/output, calculating units functioning in parallel and connected, in which each connection has its own weight and the function is determined by the structure of the network [9, 24]. In this study, we use the *Multilayer Perceptron* (MLP) algorithm.
3. K-nearest neighbor (KNN): the main idea of this algorithm is to take decisions by seeking similar cases already saved in memory [9].
4. Rule-based classifier: an associated rule is expressed by the form IF THEN. In this study, we use the *One Rule* (One-R) that generates one rule for each predictor in the data, and then selects the rule with the smallest error as its "one rule" [12].
5. Random Forest (RF): It combines the predictions made by multiple decision trees, where each decision tree is constructed from a random subset of the training dataset [11, 13].
6. Support vector machine (SVM): it is a classification technique of both linear and nonlinear data which allows constructing hyperplanes in a multidimensional space that separate cases of different class labels [9, 10].

3   Experimental Study: Application to Osteoporosis in Tunisian Population

In this section, we present the experimental methodology, the data set and the obtained results. Our experimental study is divided in two parts. In the first part, we compare the performances of different supervised classification algorithms for predicting osteoporosis in order to identify the best algorithms with the optimum classification parameters. In the second part, we use the C4.5 algorithm to extract the most important risk factors that contribute most greatly to the disease progression.

3.1   Description of the Data Set

The osteoporosis data obtained from a survey carried in La Rabta hospital in Tunisia by Sassi et al. [14, 15]. All patients were postmenopausal women. Table 1 shows the information about datasets. Data related to various risk factors were stored for each patient including polymorphisms of three candidate genes (*Vitamine D receptor* (VDR), *Low-density lipoprotein receptor-related protein 5* (LRP5) and *Receptor activator of nuclear factor kappa-B ligand* (RANKL)). The last attribute represents the class (normal, osteopenia, and osteoporosis).

- Number of instances 566 (postmenopausal women)
- Number of attributes: 29 plus the class attribute
- Each instance has one of 3 possible classes: normal (N), osteopenia (OP) or osteoporosis (O)
- Class distribution: Normal: 231, Osteopenia: 194, Osteoporosis: 141.

Diagnosis of osteoporosis is based on BMD measure, according to the World Health Organisation criteria, women were classified as Osteoporotic (T-score ≤ -2.5), Osteopenia ($-2.5 <$ T-score ≤ -1) and as Normal (T-score > -1) [1]. However, this measure can be an overlap between osteoporosis and osteopenia people [8].

We decide to remove attributes related to BMD which is a determinant factor in order to better identifying other risk factors influences.

3.2   Concepts and Experimental Protocol to Compare the Performances of Classification Algorithms for Osteoporosis Prediction

In order to categorize subjects into three classes (normal, osteopenia, osteoporosis), six classification techniques are tested in this paper. In addition, our aim is to compare the efficiency and the performance of each classifier in order to identify the best classifier with the optimal parameters. All experiments are carried out using the Weka (Waikato Environment for Knowledge Analysis) tool [16].

Test Method for Classification. We use 10-folds cross-validation as a test method. Cross-validation is a statistical methods to improve the reliability of estimators and statistical tests, especially for small database. 10 folds cross validation works as follow: first, the dataset is randomly divided into ten distinct parts. Then, the model is trained with nine of the data and tested by the tenth of data in order to predict its performance. This process is then repeated nine more times [1]. Each time, a different tenth of the data was used as a test data, and a different nine-tenths of the data were used as a training data [1].

Performance Indicator. For comparative study, we use the following performance indicators:

Percentage of correctly classified (PCC): Well classified objects/Total number of objects
Receiver Operating Characteristic curve (ROC): allows comparing supervised learning algorithms independently of the distribution of modalities of variables to predict in the test file [11].

Table 1. The dataset description

Attributes	Domain	Attributes	Domain
LRP5 gene (LRP5.1)	[G/A, G/G, A/A)	Right Hip BMD (bmd-rh) (g/cm^2)	[0.468, 1.268]
LRP5 gene (LRP5.2)	[T/C, C/C, T/T]	Right Hip T-score (Tscorerh)	[−4.1, 2.8]
LRP5 genotype for both markers	[GATC, GGTC, GGCC, AATT, GATT, GACC, GGTT]	Left Hip BMD (bmd-lh) (g/cm^2)	[0.593, 1.977]
VDR gene (ApaI)	[T/T, G/T, G/G]	Left Hip T-score (Tscorelh)	[−3.2, 2.5]
VDRgene (TaqI)	[T/C, T/T, C/C]	Total Hip BMD (bmdtoth) (g/cm^2)	[0.593, 1.977]
VDR genotype for both markers	[TTTC, GTTC, GTTT, TTTT, GGCC, GGTC, GTCC, TTCC, GGTT]	T-score of hib total (tscor_htot)	[−3.2, 2.2]
RANKL gene (693G > C)	[CC, GC, GG]	Age (ans)	[42, 81]
RANKL gene (643 C > T)	[CC, CT, TT]	Menarche age (ans)	[9, 19]
RANKLgenotype for both markers	[CCCC, CCCT, GCCC, GGCC, GCCT, CCTT, GGCT, GGTT, GCTT]	Menopause age (ans)	[42, 62]
Weight (kg)	[42, 123]	Parity	[0, 11]
Height (cm)	[141, 180]	Fracture	[0,1]
BMI (Kg/cm^2)[a]	[18.97, 45.77]	Fracture type	[FP, VP]
Spine BMD(g/cm^2)	[0.550, 1.450]	Calcium-intake (mg/j)	[247, 754]
Spine BMD_corrected (g/cm^2)	[0.531, 1.375]	Physical activity	[1.5, 10.87]
Spine T-score	[−4.8, 3.4]	**Class**	**[O, OP, N]**

[a]Body mass index (BMI)

Precision: is a measure of correctness. It's the number of objects correctly classified divided by the total number of objects labeled as belonging to the positive class. It is usually expressed as a percentage. Precision = TP/(TP + FP) [11].

Recall is the number of true positives divided by the total number of objects that actually belong to the positive class. Recall = TP/(TP + FN) [11].

F-Measure is a combined measure for precision and recall calculated as:
F-Measure = 2*Precision*Recall/(Precision + Recall) [11].

Parameters Variation of Algorithms. We attempt to explore the variation of parameters for each algorithm in order to find the optimum parameters. This part summarizes the parameters used for each algorithm.

- For C4.5 algorithm, three parameters should vary that can influence results:
  - *Confidence factor* (CF) which sets the confidence threshold for the pruning procedure ranging from {0.1 to 0.5} [17, 18]. We take three values (CF = 0.1, CF = 0.25, CF = 0.5)
  - *Minimum number of objects* (MinNumObj) which controls the minimum number of instances per leave, in the range {1, ...} [1, 17, 18]. We take three values (MinNumObj (MinNumObj = 1, MinNumObj = 2, MinNumObj = 5)
  - *Unpruned* = false, *unpruned* = true
    We use the default setting value for the remaining parameters because it can achieve the best results in most cases.
- For MIP algorithm, three parameters should vary that can influence results:
  - *Learning Rate*, ranging from {0 to 1} [19]. We take three values (0.1, 0.2, 0.3)
  - *Momentum* which helps the back propagation algorithm to get out of local minima. This parameter ranges from {0 to 1} [19]. We take three values (momentum = 0.1, momentum = 0.2, momentum = 0.6)
  - *Hidden layer* ranging from {1,2,3...} [1]. We take three values of hidden layers (2, 4, 8)
- For KNN algorithm, it is sensitive to the K value defined by the user. We vary this value to show its impact on results. The K value must be in the range of {1 to 10} [20]. We take (K = 1, k = 5, k = 10). The default WEKA parameters are used for the other parameters.
- The RF algorithm is characterized by its number of trees. So, we vary the parameters '*numtrees*' to show its impact on results. Typically the number of trees is 10 or 30 or 100 [12]. For the other parameters we use the default value because it can achieve a better result in most cases.
- For the SVM, two parameters are considered:
  - The value of *constant c* which controls the tradeoff between fitting the training data and minimizing the separating margin. This value should be in the range from {0.01 to 100} [21, 22]. For our experiment we choose (c = 0.01, c = 1, c = 10, c = 100).
  - The choice of the *kernel* function: *polynomial kernel* or *RBF kernel.*

Data Pre-processing. During this phase we have make the following tasks:

1. Applications of *Normalization* filter for all algorithms.
2. Applications of *Discredited* filter for MIP and One-R algorithms.
3. Application of *Nominal to binary* filter for the KNN algorithm.

Experimental Results Using Cross Validation Test. The 10- folds cross-validation test mode is selected for the experiments. The obtained results are showed in Tables 2, 3, 4, 5, 6 and 7.

Table 2. Evaluation of OneR algorithm using 10 folds cross-validation

PCC (%)	Precision (%)	ROC	Recall (%)	F-measure (%)
44.16	42.9	0.56	44.2	42.1

Table 3. Evaluation of KNN algorithm using 10 folds cross-validation

K	PCC (%)	Precision (%)	ROC	Recall (%)	F-measure (%)
1	41.15	41.3	0.54	41.2	41.1
5	41.16	41.5	0.57	41.2	41.3
10	45.58	46.1	0.58	45.6	44.9

Table 4. Evaluation of RF algorithm using 10 folds cross-validation

Numtree	PCC (%)	Precision (%)	ROC	Recall (%)	F-measure (%)
10	50.88	51.2	0.68	50.9	50.9
30	52.47	52.6	0.69	52.5	52.3
100	52.82	52.9	0.69	52.8	52.6

Table 5. Evaluation of MLP algorithm using 10 folds cross-validation

LearningRate	Momentum	Hiddenlayers	PCC (%)	Precision (%)	ROC	Recall (%)	F_measure (%)
0.1	0.1	2	53.88	55	0.67	53.9	54.2
0.1	0.1	4	50.17	50.9	0.66	50.2	50.8
0.1	0.1	8	51.23	51.6	0.65	51.2	51.4
0.1	0.2	2	53.71	54.8	0.68	53.7	54
0.1	0.2	4	50	50.8	0.66	50	50.2
0.1	0.2	8	49.47	50.3	0.64	49.5	49.8
0.1	0.6	2	53.53	54.6	0.67	53.5	53.8
0.1	0.6	4	49.82	50.8	0.64	49.8	50.1
0.1	0.6	8	50.17	50.3	0.66	50.2	50.2
0.2	0.1	2	52.29	53.6	0.67	52.3	52.6
0.2	0.1	4	49.64	49.8	0.64	49.6	49.7
0.2	0.1	8	48.23	47.8	0.64	48.2	48
0.2	0.2	2	50.35	51.1	0.65	50.4	50.6
0.2	0.2	4	50.17	50.1	0.65	50.2	50.1
0.2	0.2	8	48.93	48.8	0.65	48.9	48.8
0.2	0.6	2	52.47	53.3	0.66	52.5	52.7
0.2	0.6	4	50.88	51.2	0.66	50.9	51
0.2	0.6	8	47.70	47.4	0.65	47.7	47.5
0.3	0.1	2	54.24	54.5	0.67	54.2	54.3
0.3	0.1	4	53.53	54.2	0.66	53.5	53.7
0.3	0.1	8	48.23	49	0.64	48.2	48.5

(Continued)

Table 5. (*Continued*)

LearningRate	Momentum	Hiddenlayers	PCC (%)	Precision (%)	ROC	Recall (%)	F_measure (%)
0.3	0.2	2	49.64	50.4	0.66	49.6	49.9
0.3	0.2	4	48.58	49.5	0.66	48.6	48.9
0.3	0.2	8	47.70	47.6	0.63	47.7	47.6
0.3	0.6	2	54.06	55.3	0.67	54.1	54.4
0.3	0.6	4	52.29	53.2	0.68	52.3	52.4
0.3	0.6	8	48.93	49.4	0.66	48.9	49

Table 6. Evaluation of SVM algorithm using 10 folds cross-validation

C	Kernel function	PCC (%)	Precision (%)	ROC	Recall (%)	F-measure (%)
0.01	Polynomial kernel	40.81	16.7	0.5	40.8	23.7
0.01	RBF kernel	40.81	16.7	0.5	40.8	23.7
1	Polynomial kernel	54.06	54	0.66	54.1	53.8
1	RBF kernel	42.04	39.9	0.52	42	30.6
10	Polynomial kernel	54.59	54.7	0.67	54.6	54.5
10	RBF kernel	51.23	51.8	0.63	51.2	50.5
100	Polynomial kernel	55.65	55.5	0.68	55.7	55.4
100	RBF kernel	55.30	55.8	0.67	55.3	55.2

Table 7. Evaluation of C4.5 algorithm using 10 folds cross-validation

CF	MinNumObj	Unpruned	PCC (%)	Precision (%)	ROC	Recall (%)	F_measure (%)
0.1	1	False	52.82	52.6	0.65	52.8	52.7
0.1	2	False	52.47	52.4	0.66	52.5	52.5
0.1	5	False	51.59	51.7	0.66	51.6	51.4
0.25	1	False	52.82	52.6	0.65	52.8	52.7
0.25	2	False	52.47	52.4	0.66	52.5	52.4
0.25	5	False	51.59	51.7	0.66	51.6	51.4
0.5	1	False	52.82	52.6	0.65	52.8	52.7
0.5	2	False	52.47	52.4	0.66	52.5	52.4
0.5	5	False	51.59	51.7	0.66	51.6	51.4
0.1	1	True	48.05	48	0.6	48.1	48
0.1	2	True	49.29	49.4	0.61	49.3	49.3
0.1	5	True	51.06	51.1	0.65	51.1	51.1
0.25	1	True	48.05	48	0.6	48.1	48
0.25	2	True	49.29	49.4	0.61	49.3	49.3
0.25	5	True	51.06	51.1	0.65	51.1	51.1
0.5	1	True	48.05	48	0.6	48.1	48
0.5	2	True	49.29	49.4	0.61	49.3	49.3
0.5	5	True	51.06	51.1	0.65	51.1	51.1

242   N. Guannoni et al.

Thus, in order to compare the selected algorithms, we take the highest rate for all performance indicators. Results are shown in the following histograms (Figs. 1, 2, 3 and 4).

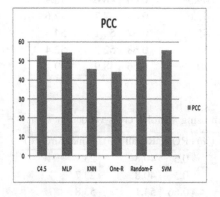

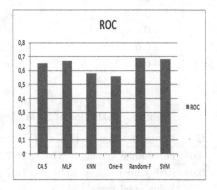

Fig. 1. PCC results for selected classification   **Fig. 2.** ROC results for selected classification

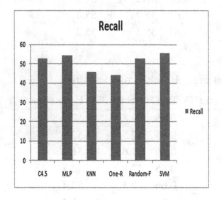

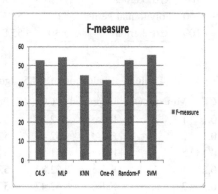

Fig. 3. Recall   results   for   selected   **Fig. 4.** F-measure results for selected classification

3.3   Concepts and Experimental Protocol to Extract the Most Important Risk Factors for Osteoporosis Occurrence

In this part, we focus to find the most important factors that have a significant bearing on the onset of osteoporosis. Those selective risk factors can be used to offer people with early prevention for helping diagnosis, monitoring disease progression, and many others.

Figure 5 represents a model consisting of nodes and branches, each attribute represents a risk factor of osteoporosis, each value is for example (>0 or <=0) depending on the existence or not of a fracture risk factor and each terminal node represent a class label (O: osteoporosis, OP: osteopenia, N: Normal bone).

The most risk factors found by C4.5 are *fracture, physical activity* <= 3.49, *low calcium intake, weight* <= 58 and *parity* > 1. VDR and RANKL genes are also associated with osteoporosis risk.

Note that the values within each branch are normalized (*within* range [0, 1]).

4   Discussion

We start by discuss the performance results of classification algorithms to identify the best algorithms with the optimum classification parameters values.

4.1   Experimental Discussion of Classification Results

This study presents the assessment of six classifiers in a small dataset of 566 instances (141 osteoporosis, 194 osteopenia and 231 healthy cases) to predict osteoporosis in postmenopausal Tunisian women based on twenty clinical factors.

- As showing in Figs. 1, 2, 3 and 4, the SVM classifier gives the best result among the six classifiers used. It performs better than the other classification techniques. It has a higher precision (55.5 %), a higher percentage of correctly classified instances (55.65 %), the best recall value (55.7 %) and the best F-measure (55.4) with C = 100, kernel = *polykernal*. So, the constant C value and the kernel type represent the indispensable parameters in the classification results. Yet, when we increase the C value, the precision rate increased.
- MLP also performed well as compared to other machine learning methods and gives a better precision, a higher percentage of correctly classified instances (54.24 %) and a ROC area value (0.54) with *learning Rate* = 0.3, *Momentum* = 0.1 and *hidden layers* = 2.

The C4.5 and the RF seem to have similar results on average. There are no significant differences between the two classifiers. The C4.5 gives a higher PCC (52.82 %) with CF = [0.1 to 0.5], *MinNumObj* = 1 and with pruning. The RF gives a higher PCC (52.82 %), ROC (0.69), Recall and F-measure of around 0.52 with *numtree* = 100.

But we observe that when we apply the unpruned = true in C4.5 algorithm, the result considerably reduced and the error rate increased. This means that pruning can reduces the size of decision trees and gives higher results of accuracy.

For the RF classifier, the best way is to choose the number of tree = 100 because it is the best value used in [23]. The worst performance can be seen to the One-R classifier and KNN classifier where the accuracy was less than 50 %.

- In general, SVM, ANN and RF are performing well as compared to other machine learning methods for prediction the bone status in a Tunisian population. In addition, the performance of these algorithms is dependent on the optimum parameters values that we have chosen.

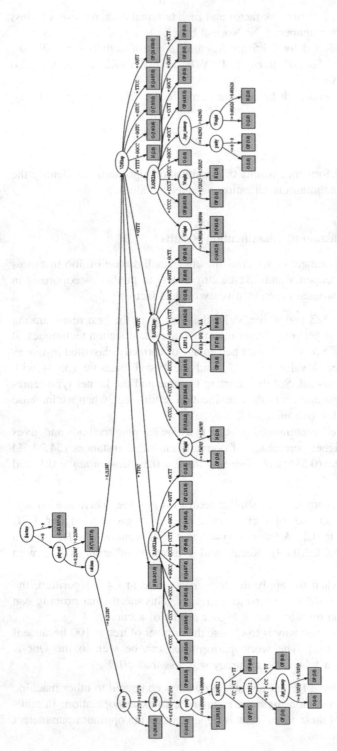

Fig. 5. Pruned Classification Tree for postmenopausal osteoporosis in Tunisia population

These examples are learning poorly since incorrectly classified instances rate is far greater. This can be explained by two reasons. On one hand, the data set used in this experiment is small (the number of women included in the test was too low). On the other hand, the BMD factors are not included in this study which is an important determinant of osteoporosis. Despite we have removed these factors, we found acceptable results with the remaining risk factors, using SVM algorithm.

4.2   Experimental Discussion of Risk Factors Selection

Figure 5 shows that the most important risk factors associated with the osteoporosis development in Tunisian postmenopausal women, are the following, in order of importance: *fractures, physical activity <= 3.49, low calcium intake, Weight <= 58* and *Parity > 1*

Despite the fact that age is a very important factor in the occurrence of osteoporosis in women, given that all surveyed women were postmenopausal in our study, in this case, age is not an important factor. In addition, VDR gene is associated with osteoporosis development when combined with a low calcium intake and a low physical activity. RANKL gene is also associated with osteoporosis risk when combinated with the VDR gene (TTTT, GTTC, GTTT, GGTC), a low *calcium intake* and a low *physical activity*.

Menopause age is a risk factor of osteoporosis but it is not significant.

Studies by biologists [14, 15], have used a statistic classical methods such as $\chi 2$-test followed by odds ratio for association measures and the logistic regression which characterized the relationship between attributes of the data set in order to extract the most important risk factors associated to the osteoporosis development. All statistical analysis used in [14, 15] were examined by SPSS software. They have found that age > 55, BMI < 27, parity > 5, calcium Intake, physical activity $\leq$ 3.5 and menopause age are the most important risk factors associated with the osteoporosis development. For genes association, Sassi et al. [14, 15], have found that VDR and RANKL genes are associated with the osteoporosis development. LRP5 gene polymorphism is not implicated in osteoporosis onset.

For osteopenia, classical methods of biologists show that *physical activity* < 3.5, a *calcium intake* < 400, *age* > 55 and *BMI* < 27 are the most important risk factors associated to the osteopenia development. VDR gene and RANL are also involved in osteopenia. LRP5 gene polymorphism is at risk of developing osteopenia when it is associated with a *parity* > 5. For the other risk factors, no effect has been found. Whereas, the C4.5 algorithm shows that *physical activity* < 3.49, *calcium intake* <= 368, *parity* > 1 and *menopause age* are associated to the osteopenia development. VDR and genes are also associated to the osteopenia development using the decision tree algorithm. This is in agreement with the study of Sassi et al. [14, 15].

The obtained results by C4.5 is almost similar with those for medical studies conducted, except for the age and BMI which are not significant using the decision tree algorithm. But the main advantage of this classification technique is the possibility to analyze risk factors jointly, and this allows the importance of each to be established towards osteoporosis or osteopenia. Another advantage that is even without accounting for bone mineral density which is the important determinant of osteoporosis, we found interesting results. We can conclude that C4.5 algorithm is more efficient in term of rapidity and accuracy compared to classical and statistical methods used by biologists.

In addition, although we have a partial database (lack of real biochemical data, lack of associated genes, low sample size, etc.), such classification algorithms provide performance that exceeds 50 %.

So, based on a rich database, the physicians can use such classification algorithms to enhance decision making.

5   Conclusion and Future Works

In conclusion, we note that it is not possible to compare the classifier' performance results of this study with the literature and declare the better. This is due to different data sets used, as well as due to different risk factors and candidature genes that are transmitted to the classifiers.

In this paper, we have presented a comparative study between classification algorithms in order to predict osteoporosis in Tunisian population. The results showed that SVM and ANN classifiers are superior to the other tested algorithms. On the other hand, we have used the *decision tree* classifier C4.5 to extract the most important risk factors for osteoporosis occurrence. The results showed that C4.5 algorithm performs best as compared to other classical methods used by biologists.

A noted limitation of this study is that, due to the small sample size, the number of women included in the test was too low to obtain good dataset classifications. Moreover, the diagnosis accuracy is still not satisfying. So, a dataset based on a larger sample size and more advanced methods are needed for improved accuracy. Furthermore, we have not studied the influence of feature selection methods on the classifiers performances, which would have been interesting for our study.

Another limit is related to the fact that classification algorithms require a good setting of parameters that is a complex and inevitable issue.

Therefore, in a future work, we would like to improve these results by creating an enhanced or updated dataset and make a follow up of women who have osteopenia and who progress to osteoporosis to better predict which women can become osteoporotic and which others can remain with osteopenia. In addition, we would like to apply feature selection methods in order to improve the classification results. After this, we try to develop new powerful classification algorithms (based on times series, Bayesian relief Network, SVM) to build a comprehensive model that can guide medical decision-making and that can handle various potential risk factors simultaneously.

References

1. Iliou, T., Anagnostopoulos, C.N., Anastassopoulos, G.: Osteoporosis detection using machine learning techniques and feature selection. Int. J. Artif. Intell. Tools **23**(05), 1450014 (2014)
2. Masood, Z., Shahzad, S., Saqib, A., Khizer, A.: Osteopenia and osteoporosis; frequency among females. Prof. Med. J. **21**, 477 (2014)
3. International osteoporosis foundation (IOF). http://www.iofbonehealth.org/epidemiology
4. Kim, S.K., Yoo, T.K., Kim, D.W: Osteoporosis risk prediction using machine learning and conventional methods. In: Engineering in Medicine and Biology Society (EMBC 35th Annual International Conference of the IEEE), pp. 188–191 (2013)
5. Younesi, E.: A knowledge-based integrative modeling approach for in-silico identification of mechanistic targets in neurodegeneration with focus on Alzheimer's disease. Ph.D., Universitäts-und Landesbibliothek Bonn (2014)
6. Chang, H.W., Chiu, Y.H., Kao, H.Y., Yang, C.H., Ho, W.H: Comparison of classification algorithms with wrapper-based feature selection for predicting osteoporosis outcome based on genetic factors in a taiwanese women population. Int. J. Endocrinol. (2013)
7. Yoo, T.K., Kim, S.K., Oh, E., Kim, D.W.: Risk prediction of femoral neck osteoporosis using machine learning and conventional methods. In: Rojas, I., Joya, G., Cabestany, J. (eds.) IWANN 2013, Part II. LNCS, vol. 7903, pp. 181–188. Springer, Heidelberg (2013)
8. Xu, Y., Li, D., Chen, Q., Fan, Y.: Full supervised learning for osteoporosis diagnosis using micro-CT images. Microsc. Res. Tech. **76**(4), 333–341 (2013)
9. Han, J., Kamber, M., Pei, J.: Data Mining: Concepts and Techniques: Concepts and Techniques. Elsevier, New York (2011)
10. Mhamdi, F., Elloumi, M.: A new survey on knowledge discovery and data mining. In: RCIS, pp. 427–432 (2008)
11. Aleem, S., Capretz, L.F., Ahmed, F: Benchmarking machine learning technologies for software defect detection (2015). arXiv preprint arXiv:1506.07563
12. Ross, P.: Data mining. http://www.soc.napier.ac.uk/∼peter/vldb/dm/node8.html
13. Moudani, W., Shahin, A., Chakik, F., Rajab, D.: Intelligent predictive osteoporosis system. Int. J. Comput. Appl. (IJCA) **32**(5), 28–37 (2011)
14. Sassi, R., Sahli, H., Souissi, C., Sellami, S., Ben Ammar El Gaaied, A.: Polymorphisms in VDR gene in Tunisian postmenopausal women are associated with osteopenia phenotype. Climacteric **18**, 624–630 (2015)
15. Sassi, R., Sahli, H., Souissi, C., El Mahmoudi, H., Zouari, B., ElGaaied, A.B.A., Ferrari, S.L.: Association of LRP5 genotypes with osteoporosis in Tunisian post-menopausal women. BMC Musculoskeletal Disorders **15**(1), 144 (2014)
16. Markov, Z., Russell, I.: An introduction to the weka data mining system. In: ACM SIGCSE Bulletin, vol. 38, pp. 367–368. ACM (2006)
17. Optimizing parameters. The University of Waikato
18. Koblar, V.: Optimizing parameters for machine learning algorithm. Technical report, Jozef Stefan International postgraduate school (2012)
19. Wu, C.H., McLarty, J.W. (eds.): Neural networks and genome informatics, 1st edn. Elsevier, Amsterdam (2012)
20. Baskin, I., Tetko, I., Varnek, A.: Tutorial on machine learning, Part 1. In: Benchmarking of Different Machine Learning Regression Methods. http://infochim.ustrasbg.fr/CS3/program/Tutorials/Tutorial2a.pdf. (last visited January 21, 2016)

21. Ferrier, J.L., Bernard, A., Gusikhin, O., Madani, K. (eds.): Selected Papers from the International Conference on Informatics in Control Automation and Robotics 2006, vol. 283. Springer, Heidelberg (2014)
22. Batsaikan, O., Ho, C.K., Singh, Y.P.: A genetic algorithm-based multi-class support vector machine for mongolian character recognition. INFOCOMP J. Comput. Sci. **8**(1), 1–7 (2009)
23. Wang, Q., Zhang, L., Chi, M., Guo, J.: MTForest: ensemble decision trees based on multi-task learning. In: ECAI, pp. 122–126 (2008)
24. Amrani, M.: Surveillance et diagnostic d'une ligne de production par les réseaux de neurones artificiels. Ph.d., Université M'hamedBougara de Boumerdès (2010)

Process Mining: Towards Comparability of Healthcare Processes

Emmanuel Helm[1(✉)] and Josef Küng[2]

[1] Research Department of e-Health, Integrated Care,
University of Applied Sciences Upper Austria, 4232 Hagenberg, Austria
emmanuel.helm@fh-hagenberg.at
[2] Institute for Applied Knowledge Processing,
Johannes Kepler University, 4040 Linz, Austria
jkueng@faw.jku.at
https://www.fh-ooe.at/
http://www.jku.at/

Abstract. With the technology emerging more and more possible applications of process mining in healthcare become apparent. In most cases the goal of applying process mining to the healthcare domain is to find out what actually happened and to deliver a concise assessment of the organizational reality by mining the event logs of health information systems. To develop medical guidelines or patient pathways considering economic aspects and quality of care, a comparative analysis of different existing approaches is useful (e.g. how different hospitals execute the same process in different ways). This work discusses how to use existing process mining techniques for comparative analysis of healthcare processes and presents an approach based on the L* life-cycle model.

Keywords: Process mining · Data mining · Process quality

1 Motivation

One reason for the lack of Business Process Management (BPM) technologies in healthcare is the complexity of the processes, where unforeseen events in the course of a disease or during the treatment are to some degree a "normal" phenomenon [1].

Process mining provides an a-posteriori empirical method to *discover* processes in observed system behavior (i.e. event logs) [2]. A goal of applying process mining techniques to the healthcare domain is to understand the complex interactions between multiple actors, both human and machine, and the underlying, partially implicit processes [3].

To develop medical guidelines or patient pathways considering economic aspects and quality of care, a comparative analysis of different existing approaches is useful. Partington et al. propose the application of process mining as *an evidence-based business process analysis method* to investigate variations in clinical practice and delivery of care across different hospital settings [4].

© Springer International Publishing Switzerland 2016
M.E. Renda et al. (Eds.): ITBAM 2016, LNCS 9832, pp. 249–252, 2016.
DOI: 10.1007/978-3-319-43949-5_20

2   Problem

The characteristics of healthcare processes make it impossible to apply rigorous BPM, Workflow Management (WFM) and Business Process Reengineering (BPR) techniques. Mans et al. make it clear that *a hospital is not a factory and patients cannot be cured using a conveyor belt system* [5]. However, the authors of [3,5,6] (among others) agree that process mining has the potential to improve healthcare processes by increasing compliance and performance while reducing costs.

The comparison of mined healthcare processes aims to show the (dis)similarity of practices across different healthcare providers and to identify potential improvements. In addition to the general challenges of applying process mining techniques to the healthcare domain, comparative analysis also has to deal with the *gaps* between different healthcare providers. These gaps mostly originate from the fact that different organizations are essentially executing the same process without following a strict process model [7].

To enable the comparability of two mined process models, shared semantics are necessary (i.e. using the same terms for the same activities and characteristics). The precondition for semantic interoperability is a formal representation of data within the healthcare information systems. Since healthcare systems are often heterogeneous and autonomous IT systems, the formal representation varies strongly [6].

Only two approaches that compare the processes of different healthcare providers were found in the literature. While Partington et al. [4] actually compared data from different sources (i.e. different information systems), for Mans et al. [8] the basis was a shared database, filled by different hospitals.

3   Approach

The presented approach extends the L^* *life-cycle model* for process mining to support comparative analysis and cross-organizational mining [7]. It is based on a case study comparing four hospitals (cf. [4]) and on the experience gathered during a process mining project comparing eight Austrian hospitals.

The critical stages of *extraction* (1), *control-flow model creation* (2) and *model enhancement* (3) [7] are extended to allow for parallel execution, thus enabling interaction between the different mining activities (i.e. between the mining of processes from different hospitals).

4   Results

Figure 1 shows the extended L* life-cycle model for comparative analysis. It comprises all stages of the original model but spares the steps between the stages for better readability (e.g. inclusion of historical data and handmade models). Previous research aimed at *Stage 1* to prepare the logs of different information systems for further analysis [9,10].

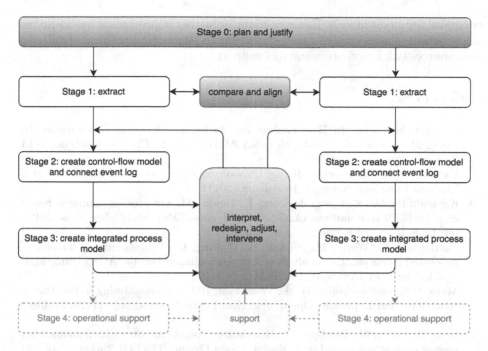

Fig. 1. The extended *L* life-cycle model* for comparative analysis, based on the original model in [7] and the adoption in [4]. Continuous alignment between the parallel stages is necessary to minimize the number of necessary iterations.

On the left and right side in Fig. 1 the two parallel mining processes comprising the respective stages are depicted. Extending the original L* life-cycle model, the *interpretation, intervention, adjustment* and *redesign* steps are conducted with both models together. Additionally a new step during the *extraction* stage was added, to *compare and align* the logs before applying automated process discovery techniques.

Currently methods are developed to show possible gaps at all stages. After identifying a gap, key figures indicate if hospitals either do fundamentally different things or they record the same things differently (e.g. using different coding systems). The mining activities can then be adopted accordingly, leading for example to further preprocessing of the logs.

The first approaches are based on statistical analysis of the base logs (e.g. t-tests based on the frequency of specific events). Further approaches will include graph similarity measurements and conformance checking techniques (cf. for example the works of Dijkman et al. [11] and Van der Aalst [12]).

5   Conclusions

By coordinating multiple mining activities it is possible to identify the gaps in early stages, thus reducing the number of iterations necessary to present meaningful, comparable process models. However, room for further improvement was identified since the early stages involving the comparison and semantic alignment of different data sources lack automation and tool support.

References

1. Reichert, M.: What BPM technology can do for healthcare process support. In: Peleg, M., Lavrač, N., Combi, C. (eds.) AIME 2011. LNCS, vol. 6747, pp. 2–13. Springer, Heidelberg (2011)
2. Van der Aalst, W.: Process Mining: Discovery, Conformance and Enhancement of Business Processes. Springer, Heidelberg (2011)
3. Kaymak, U., Mans, R., van de Steeg, T., Dierks, M.: On process mining in health care. In: IEEE International Conference on Systems, Man, and Cybernetics (SMC), pp. 1859–1864. IEEE (2012)
4. Partington, A., Wynn, M., Suriadi, S., Ouyang, C.: Process mining for clinical processes: a comparative analysis of four Australian hospitals. ACM Trans. Manage. Inf. Syst. (TMIS) **5**(4), 19 (2015)
5. Mans, R.S., van der Aalst, W.M., Vanwersch, R.J.: Process Mining in Healthcare: Evaluating and Exploiting Operational Healthcare Processes, pp. 17–26. Springer, Heidelberg (2015)
6. Lenz, R., Peleg, M., Reichert, M.: Healthcare process support: achievements, challenges, current research. Int. J. Knowl. Based Organ. (IJKBO) **2**(4), 1–16 (2012)
7. Van der Aalst, W., et al.: Process mining manifesto. In: Daniel, F., Barkaoui, K., Dustdar, S. (eds.) BPM Workshops 2011, Part I. LNBIP, vol. 99, pp. 169–194. Springer, Heidelberg (2012)
8. Mans, R., Schonenberg, H., Leonardi, G., Panzarasa, S., Cavallini, A., Quaglini, S., van der Aalst, W.: Process mining techniques: an application to stroke care. Stud. Health Technol. Inf. **136**, 573 (2008)
9. Paster, F., Helm, E.: From IHE audit trails to XES event logs facilitating process mining. In: Digital Healthcare Empowering Europeans: Proceedings of MIE2015, vol. 210, p. 40 (2015)
10. Helm, E., Paster, F.: First steps towards process mining in distributed health information systems. Int. J. Electron. Telecommun. **61**(2), 137–142 (2015)
11. Dijkman, R., Dumas, M., García-Bañuelos, L.: Graph matching algorithms for business process model similarity search. In: Dayal, U., Eder, J., Koehler, J., Reijers, H.A. (eds.) BPM 2009. LNCS, vol. 5701, pp. 48–63. Springer, Heidelberg (2009)
12. Van der Aalst, W.M.: Business alignment: using process mining as a tool for Delta analysis and conformance testing. Requirements Eng. **10**(3), 198–211 (2005)

Author Index

Printed in the United States
By Bookmasters